REGULATIONS FOR THE SAFE TRANSPORT OF RADIOACTIVE MATERIAL

2025 Edition

The following States are Members of the International Atomic Energy Agency:

AFGHANISTAN
ALBANIA
ALGERIA
ANGOLA
ANTIGUA AND BARBUDA
ARGENTINA
ARMENIA
AUSTRALIA
AUSTRIA
AZERBAIJAN
BAHAMAS, THE
BAHRAIN
BANGLADESH
BARBADOS
BELARUS
BELGIUM
BELIZE
BENIN
BOLIVIA, PLURINATIONAL STATE OF
BOSNIA AND HERZEGOVINA
BOTSWANA
BRAZIL
BRUNEI DARUSSALAM
BULGARIA
BURKINA FASO
BURUNDI
CABO VERDE
CAMBODIA
CAMEROON
CANADA
CENTRAL AFRICAN REPUBLIC
CHAD
CHILE
CHINA
COLOMBIA
COMOROS
CONGO
COOK ISLANDS
COSTA RICA
CÔTE D'IVOIRE
CROATIA
CUBA
CYPRUS
CZECH REPUBLIC
DEMOCRATIC REPUBLIC OF THE CONGO
DENMARK
DJIBOUTI
DOMINICA
DOMINICAN REPUBLIC
ECUADOR
EGYPT
EL SALVADOR
ERITREA
ESTONIA
ESWATINI
ETHIOPIA
FIJI
FINLAND
FRANCE
GABON
GAMBIA, THE
GEORGIA
GERMANY
GHANA
GREECE
GRENADA
GUATEMALA
GUINEA
GUYANA
HAITI
HOLY SEE
HONDURAS
HUNGARY
ICELAND
INDIA
INDONESIA
IRAN, ISLAMIC REPUBLIC OF
IRAQ
IRELAND
ISRAEL
ITALY
JAMAICA
JAPAN
JORDAN
KAZAKHSTAN
KENYA
KOREA, REPUBLIC OF
KUWAIT
KYRGYZSTAN
LAO PEOPLE'S DEMOCRATIC REPUBLIC
LATVIA
LEBANON
LESOTHO
LIBERIA
LIBYA
LIECHTENSTEIN
LITHUANIA
LUXEMBOURG
MADAGASCAR
MALAWI
MALAYSIA
MALDIVES
MALI
MALTA
MARSHALL ISLANDS
MAURITANIA
MAURITIUS
MEXICO
MONACO
MONGOLIA
MONTENEGRO
MOROCCO
MOZAMBIQUE
MYANMAR
NAMIBIA
NEPAL
NETHERLANDS, KINGDOM OF THE
NEW ZEALAND
NICARAGUA
NIGER
NIGERIA
NORTH MACEDONIA
NORWAY
OMAN
PAKISTAN
PALAU
PANAMA
PAPUA NEW GUINEA
PARAGUAY
PERU
PHILIPPINES
POLAND
PORTUGAL
QATAR
REPUBLIC OF MOLDOVA
ROMANIA
RUSSIAN FEDERATION
RWANDA
SAINT KITTS AND NEVIS
SAINT LUCIA
SAINT VINCENT AND THE GRENADINES
SAMOA
SAN MARINO
SAUDI ARABIA
SENEGAL
SERBIA
SEYCHELLES
SIERRA LEONE
SINGAPORE
SLOVAKIA
SLOVENIA
SOMALIA
SOUTH AFRICA
SPAIN
SRI LANKA
SUDAN
SWEDEN
SWITZERLAND
SYRIAN ARAB REPUBLIC
TAJIKISTAN
THAILAND
TOGO
TONGA
TRINIDAD AND TOBAGO
TUNISIA
TÜRKİYE
TURKMENISTAN
UGANDA
UKRAINE
UNITED ARAB EMIRATES
UNITED KINGDOM OF GREAT BRITAIN AND NORTHERN IRELAND
UNITED REPUBLIC OF TANZANIA
UNITED STATES OF AMERICA
URUGUAY
UZBEKISTAN
VANUATU
VENEZUELA, BOLIVARIAN REPUBLIC OF
VIET NAM
YEMEN
ZAMBIA
ZIMBABWE

The Agency's Statute was approved on 23 October 1956 by the Conference on the Statute of the IAEA held at United Nations Headquarters, New York; it entered into force on 29 July 1957. The Headquarters of the Agency are situated in Vienna. Its principal objective is "to accelerate and enlarge the contribution of atomic energy to peace, health and prosperity throughout the world".

IAEA SAFETY STANDARDS SERIES No. SSR-6 (Rev. 2)

REGULATIONS FOR THE SAFE TRANSPORT OF RADIOACTIVE MATERIAL

2025 EDITION

SPECIFIC SAFETY REQUIREMENTS

INTERNATIONAL ATOMIC ENERGY AGENCY
VIENNA, 2025

COPYRIGHT NOTICE

Publishing Section
International Atomic Energy Agency
Vienna International Centre
PO Box 100
1400 Vienna, Austria
tel.: +43 1 2600 22529 or 22530
email: sales.publications@iaea.org
www.iaea.org/publications

Printed by the IAEA in Austria
December 2025
STI/PUB/2120
https://doi.org/10.61092/iaea.439y-2h8h

IAEA Library Cataloguing in Publication Data

Names: International Atomic Energy Agency.
Title: Regulations for the safe transport of radioactive material : 2025 edition / International Atomic Energy Agency.
Description: Vienna : International Atomic Energy Agency, 2025. | Series: Safety standards series, ISSN 1020–525X ; no. SSR-6 (Rev. 2) | Includes bibliographical references.
Identifiers: IAEAL 25-01791 | ISBN 978-92-0-115825-3 (paperback : alk. paper) | ISBN 978-92-0-115925-0 (pdf) | ISBN 978-92-0-116025-6 (epub)
Subjects: LCSH: Radioactive substances — Transportation. | Radioactive substances — Safety measures. | Radioactive substances — Security measures.
Classification: UDC 656.073 | STI/PUB/2120

FOREWORD

by Rafael Mariano Grossi
Director General

The IAEA's Statute authorizes it to "establish...standards of safety for protection of health and minimization of danger to life and property". These are standards that the IAEA must apply to its own operations, and that States can apply through their national regulations.

The IAEA started its safety standards programme in 1958 and there have been many developments since. As Director General, I am committed to ensuring that the IAEA maintains and improves upon this integrated, comprehensive and consistent set of up to date, user friendly and fit for purpose safety standards of high quality. Their proper application in the use of nuclear science and technology should offer a high level of protection for people and the environment across the world and provide the confidence necessary to allow for the ongoing use of nuclear technology for the benefit of all.

Safety is a national responsibility underpinned by a number of international conventions. The IAEA safety standards form a basis for these legal instruments and serve as a global reference to help parties meet their obligations. While safety standards are not legally binding on Member States, they are widely applied. They have become an indispensable reference point and a common denominator for the vast majority of Member States that have adopted these standards for use in national regulations to enhance safety in nuclear power generation, research reactors and fuel cycle facilities as well as in nuclear applications in medicine, industry, agriculture and research.

The IAEA safety standards are based on the practical experience of its Member States and produced through international consensus. The involvement of the members of the Safety Standards Committees, the Nuclear Security Guidance Committee and the Commission on Safety Standards is particularly important, and I am grateful to all those who contribute their knowledge and expertise to this endeavour.

The IAEA also uses these safety standards when it assists Member States through its review missions and advisory services. This helps Member States in the application of the standards and enables valuable experience and insight to be shared. Feedback from these missions and services, and lessons identified from events and experience in the use and application of the safety standards, are taken into account during their periodic revision.

I believe the IAEA safety standards and their application make an invaluable contribution to ensuring a high level of safety in the use of nuclear technology. I encourage all Member States to promote and apply these standards, and to work with the IAEA to uphold their quality now and in the future.

PREFACE

This publication is a revision of IAEA Safety Standards Series No. SSR-6 (Rev. 1), Regulations for the Safe Transport of Radioactive Material, 2018 Edition. The revision was undertaken by amending, adding or deleting specific paragraphs.

Paragraph numbers for new paragraphs are generated by adding an uppercase letter or the next sequential uppercase letter, as appropriate, to the preceding paragraph number. This paragraph numbering practice is not meant to imply a link between existing and newly added paragraphs. The following paragraphs have been added in the 2025 Edition: 229A, 229B, 544A, 546A, 572A, 604A, 821B.

Where a paragraph has been deleted, the paragraph number is retained together with an explanatory comment. The following paragraphs have been deleted in the 2025 Edition: 209, 524A, 567, 628, 656.

In addition, Table 2 has been deleted and its content has been incorporated in a new Appendix.

Amended paragraphs retain their original paragraph number. A table of all the changes made is available upon request to the IAEA (safety-standards.contact-point@iaea.org).

THE IAEA SAFETY STANDARDS

BACKGROUND

Radioactivity is a natural phenomenon and natural sources of radiation are features of the environment. Radiation and radioactive substances have many beneficial applications, ranging from power generation to uses in medicine, industry and agriculture. The radiation risks to workers and the public and to the environment that may arise from these applications have to be assessed and, if necessary, controlled.

Activities such as the medical uses of radiation, the operation of nuclear installations, the production, transport and use of radioactive material, and the management of radioactive waste must therefore be subject to standards of safety.

Regulating safety is a national responsibility. However, radiation risks may transcend national borders, and international cooperation serves to promote and enhance safety globally by exchanging experience and by improving capabilities to control hazards, to prevent accidents, to respond to emergencies and to mitigate any harmful consequences.

States have an obligation of diligence and duty of care, and are expected to fulfil their national and international undertakings and obligations.

International safety standards provide support for States in meeting their obligations under general principles of international law, such as those relating to environmental protection. International safety standards also promote and assure confidence in safety and facilitate international commerce and trade.

A global nuclear safety regime is in place and is being continuously improved. IAEA safety standards, which support the implementation of binding international instruments and national safety infrastructures, are a cornerstone of this global regime. The IAEA safety standards constitute a useful tool for contracting parties to assess their performance under these international conventions.

THE IAEA SAFETY STANDARDS

The status of the IAEA safety standards derives from the IAEA's Statute, which authorizes the IAEA to establish or adopt, in consultation and, where appropriate, in collaboration with the competent organs of the United Nations and with the specialized agencies concerned, standards of safety for protection of health and minimization of danger to life and property, and to provide for their application.

With a view to ensuring the protection of people and the environment from harmful effects of ionizing radiation, the IAEA safety standards establish fundamental safety principles, requirements and measures to control the radiation exposure of people and the release of radioactive material to the environment, to restrict the likelihood of events that might lead to a loss of control over a nuclear reactor core, nuclear chain reaction, radioactive source or any other source of radiation, and to mitigate the consequences of such events if they were to occur. The standards apply to facilities and activities that give rise to radiation risks, including nuclear installations, the use of radiation and radioactive sources, the transport of radioactive material and the management of radioactive waste.

Safety measures and security measures[1] have in common the aim of protecting human life and health and the environment. Safety measures and security measures must be designed and implemented in an integrated manner so that security measures do not compromise safety and safety measures do not compromise security.

The IAEA safety standards reflect an international consensus on what constitutes a high level of safety for protecting people and the environment from harmful effects of ionizing radiation. They are issued in the IAEA Safety Standards Series, which has three categories (see Fig. 1).

Safety Fundamentals

Safety Fundamentals present the fundamental safety objective and principles of protection and safety, and provide the basis for the Safety Requirements. The principles are expressed as 'must' statements.

Safety Requirements

Safety Requirements are governed by the objective and principles of the Safety Fundamentals. They establish the requirements to be met to ensure the protection of people and the environment, both now and in the future. The format and style of the Safety Requirements facilitate their use for the establishment of a national regulatory framework. Requirements are presented as 'overarching' requirements[2] in bold, followed by a number of associated requirements; all are equally important and are expressed as 'shall' statements.

Safety Guides

Safety Guides provide recommendations on how to comply with the Safety Requirements, indicating an international consensus that it is necessary to take the

---

[1] See also publications issued in the IAEA Nuclear Security Series.

[2] The IAEA Regulations for the Safe Transport of Radioactive Material do not include overarching requirements.

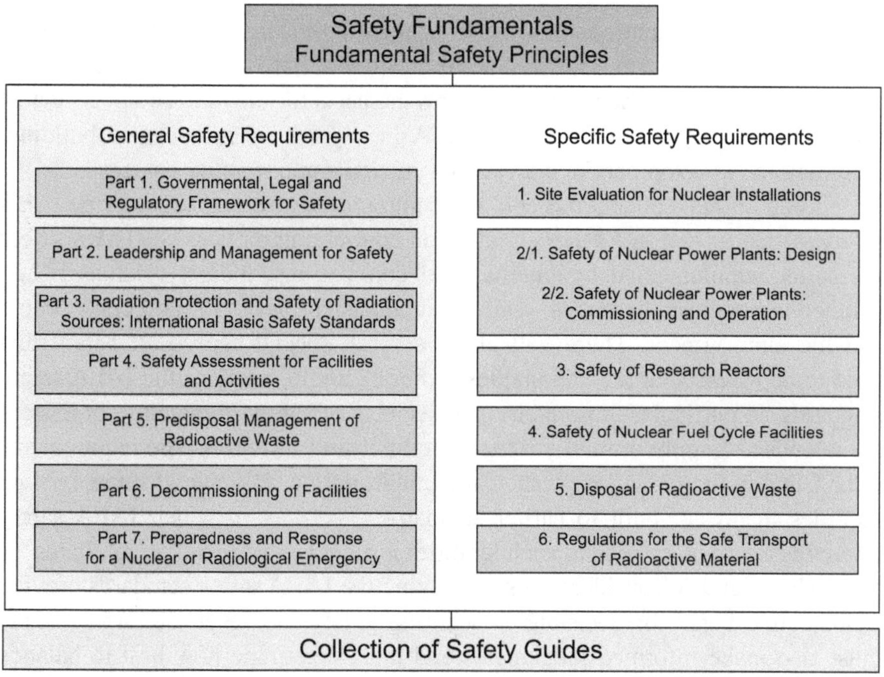

FIG. 1. The long term structure of the IAEA Safety Standards Series.

measures recommended (or alternative measures that achieve the same level of protection). Safety Guides present international good practices and, increasingly, best practices. The recommendations provided in Safety Guides are expressed as ‘should’ statements.

APPLICATION OF THE IAEA SAFETY STANDARDS

The principal users of safety standards in IAEA Member States are regulatory bodies and other relevant national authorities. The IAEA safety standards are also used by co-sponsoring organizations and by many organizations that design, construct and operate nuclear facilities, as well as organizations involved in the use of radiation and radioactive sources.

The IAEA safety standards are applicable, as relevant, throughout the entire lifetime of all facilities and activities — existing and new — utilized for peaceful purposes and to protective actions to reduce existing radiation risks. They can be used by States as a reference for their national regulations in respect of facilities and activities.

The IAEA's Statute makes the safety standards binding on the IAEA in relation to its own operations and also on States in relation to IAEA assisted operations.

The IAEA safety standards also form the basis for the IAEA's safety review services, and they are used by the IAEA in support of competence building, including the development of educational curricula and training courses.

International conventions contain requirements similar to those in the IAEA safety standards and make them binding on contracting parties. The IAEA safety standards, supplemented by international conventions, industry standards and detailed national requirements, establish a consistent basis for protecting people and the environment. There will also be some special aspects of safety that need to be assessed at the national level. For example, many of the IAEA safety standards, in particular those addressing aspects of safety in planning or design, are intended to apply primarily to new facilities and activities. The requirements established in the IAEA safety standards might not be fully met at some existing facilities that were built to earlier standards. The way in which IAEA safety standards are to be applied to such facilities is a decision for individual States.

The scientific considerations underlying the IAEA safety standards provide an objective basis for decisions concerning safety; however, decision makers must also make informed judgements and must determine how best to balance the benefits of an action or an activity against the associated radiation risks and any other detrimental impacts to which it gives rise.

DEVELOPMENT PROCESS FOR THE IAEA SAFETY STANDARDS

The preparation and review of the safety standards involves the IAEA Secretariat and five Safety Standards Committees, for emergency preparedness and response (EPReSC) (as of 2016), nuclear safety (NUSSC), radiation safety (RASSC), the safety of radioactive waste (WASSC) and the safe transport of radioactive material (TRANSSC), and a Commission on Safety Standards (CSS) which oversees the IAEA safety standards programme (see Fig. 2).

All IAEA Member States may nominate experts for the Safety Standards Committees and may provide comments on draft standards. The membership of the Commission on Safety Standards is appointed by the Director General and includes senior governmental officials having responsibility for establishing national standards.

A management system has been established for the processes of planning, developing, reviewing, revising and establishing the IAEA safety standards. It articulates the mandate of the IAEA, the vision for the future application of the safety standards, policies and strategies, and corresponding functions and responsibilities.

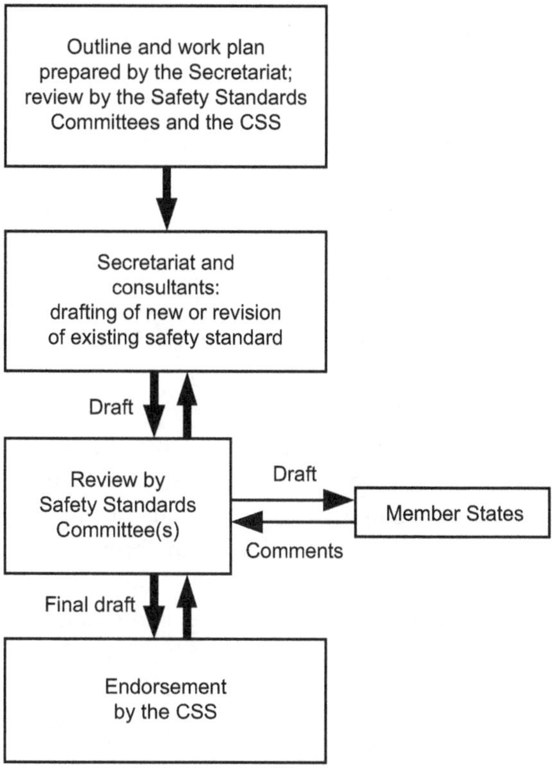

FIG. 2. The process for developing a new safety standard or revising an existing standard.

INTERACTION WITH OTHER INTERNATIONAL ORGANIZATIONS

The findings of the United Nations Scientific Committee on the Effects of Atomic Radiation (UNSCEAR) and the recommendations of international expert bodies, notably the International Commission on Radiological Protection (ICRP), are taken into account in developing the IAEA safety standards. Some safety standards are developed in cooperation with other bodies in the United Nations system or other specialized agencies, including the Food and Agriculture Organization of the United Nations, the United Nations Environment Programme, the International Labour Organization, the OECD Nuclear Energy Agency, the Pan American Health Organization and the World Health Organization.

INTERPRETATION OF THE TEXT

Safety related terms are to be understood as they appear in the IAEA Nuclear Safety and Security Glossary (see https://www.iaea.org/resources/publications/iaea-nuclear-safety-and-security-glossary). Otherwise, words are used with the spellings and meanings assigned to them in the latest edition of The Concise Oxford Dictionary. For Safety Guides, the English version of the text is the authoritative version.

The background and context of each standard in the IAEA Safety Standards Series and its objective, scope and structure are explained in Section 1, Introduction, of each publication.

Material for which there is no appropriate place in the body text (e.g. material that is subsidiary to or separate from the body text, is included in support of statements in the body text, or describes methods of calculation, procedures or limits and conditions) may be presented in appendices or annexes.

An appendix, if included, is considered to form an integral part of the safety standard. Material in an appendix has the same status as the body text, and the IAEA assumes authorship of it. Annexes and footnotes to the main text, if included, are used to provide practical examples or additional information or explanation. Annexes and footnotes are not integral parts of the main text. Annex material published by the IAEA is not necessarily issued under its authorship; material under other authorship may be presented in annexes to the safety standards. Extraneous material presented in annexes is excerpted and adapted as necessary to be generally useful.

CONTENTS

LIST OF TABLES

Section I

INTRODUCTION

BACKGROUND

101. These Regulations establish standards of safety that provide an acceptable level of control of the radiation, criticality and thermal hazards to people, property and the environment that are associated with the transport of *radioactive material*. These Regulations are based on IAEA Safety Standards Series Nos SF-1, Fundamental Safety Principles [1]; GSR Part 3, Radiation Protection and Safety of Radiation Sources: International Basic Safety Standards [2]; GSR Part 1 (Rev. 1), Governmental, Legal and Regulatory Framework for Safety [3]; GSR Part 2, Leadership and Management for Safety [4]; GSR Part 4 (Rev. 1), Safety Assessment for Facilities and Activities [5]; and GSR Part 7, Preparedness and Response for a Nuclear or Radiological Emergency [6]. Thus, compliance with these Regulations is deemed to satisfy the General Safety Requirements of the IAEA Safety Standards Series in respect of transport. In accordance with SF-1 [1], the prime responsibility for safety must rest with the person or organization responsible for facilities and activities that give rise to radiation risks.

102. These Regulations are supplemented by a number of Safety Guides, including IAEA Safety Standards Series Nos SSG-26 (Rev. 1), Advisory Material for the IAEA Regulations for the Safe Transport of Radioactive Material (2018 Edition) [7] (the 2025 edition is under development); SSG-65, Preparedness and Response for a Nuclear or Radiological Emergency Involving the Transport of Radioactive Material [8]; SSG-78, Compliance Assurance for the Safe Transport of Radioactive Material [9]; TS-G-1.4, The Management System for the Safe Transport of Radioactive Material [10]; SSG-86, Radiation Protection Programmes for the Transport of Radioactive Material [11]; SSG-33 (Rev. 1), Schedules of Provisions of the IAEA Regulations for the Safe Transport of Radioactive Material (2018 Edition) [12] (the 2025 edition is under development); and SSG-66, Format and Content of the Package Design Safety Report for the Transport of Radioactive Material [13].

103. In certain parts of these Regulations, a particular action is prescribed, but the responsibility for carrying out the action is not specifically assigned to any particular person. Such responsibility may vary according to the laws and customs of different countries and the international conventions into which these countries have entered. For the purpose of these Regulations, it is not necessary

to make this assignment, but only to identify the action itself. It remains the prerogative of each government to assign this responsibility.

OBJECTIVE

104. The objective of these Regulations is to establish requirements that must be satisfied to ensure safety and to protect people, property and the environment from harmful effects of ionizing radiation during the transport of *radioactive material*. This protection is achieved by requiring:

(a) Containment of the *radioactive contents*;
(b) Control of external *dose rate*;
(c) Prevention of criticality;
(d) Prevention of damage caused by heat.

These requirements are satisfied firstly by applying a graded approach to contents limits for *packages* and *conveyances* and to performance standards applied to *package designs*, depending upon the hazard of the *radioactive contents*. Secondly, they are satisfied by imposing conditions on the *design* and operation of *packages* and on the maintenance of *packagings*, including consideration of the nature of the *radioactive contents*. Thirdly, they are satisfied by requiring administrative controls, including, where appropriate, *approval* by *competent authorities*. Finally, further protection is provided by making arrangements for planning and preparing emergency response to protect people, property and the environment.

105. In the transport of *radioactive material*, the safety of people and the protection of property and the environment are assured when these Regulations are complied with. Confidence in this regard is achieved through a *management system* and *compliance assurance* programmes.

SCOPE

106. These Regulations apply to the transport of *radioactive material* by all modes on land, water, or in the air, including transport that is incidental to the use of the *radioactive material*. Transport comprises all operations and conditions associated with, and involved in, the movement of *radioactive material*; these include the *design*, manufacture, maintenance and repair of *packaging*, and the preparation, consigning, loading, carriage including in-transit storage,

shipment after storage, unloading and receipt at the final destination of loads of *radioactive material* and *packages*. A graded approach is applied in specifying the performance standards in these Regulations, which are characterized in terms of three general severity levels:

(a) Routine conditions of transport (incident free);
(b) Normal conditions of transport (minor mishaps);
(c) Accident conditions of transport.

107. These Regulations do not apply to any of the following:

(a) *Radioactive material* that is an integral part of the means of transport.
(b) *Radioactive material* moved within an establishment that is subject to appropriate safety regulations in force in the establishment and where the movement does not involve public roads or railways.
(c) *Radioactive material* implanted or incorporated into a person or live animal for diagnosis or treatment.
(d) *Radioactive material* in or on a person who is to be transported for medical treatment because the person has been subject to accidental or deliberate intake of *radioactive material* or to *contamination*.
(e) *Radioactive material* in consumer products that have received regulatory authorization, following their sale to the end user.
(f) *Naturally occurring radioactive material*, provided the activity concentration of the material does not exceed 10 times the values specified in the Appendix, or as calculated in accordance with paras 405 and 406. For *naturally occurring radioactive material* where the radionuclides are not in secular equilibrium, the calculation of the activity concentration shall be performed in accordance with para. 405.
(g) Non-radioactive solid objects with radioactive substances present on any surface in quantities not in excess of the levels defined in para. 214.

108. These Regulations do not specify controls such as routeing or physical protection that may be instituted for reasons other than radiation safety. Any such controls shall take into account radiological and non-radiological hazards, and shall not detract from the standards of safety that these Regulations are intended to provide.

109. Measures are expected to be taken to ensure that *radioactive material* is kept secure in transport so as to prevent unauthorized removal or damage and to ensure that control of the *radioactive material* is not relinquished inappropriately (see Annex I).

110. For *radioactive material* having other dangerous properties, and for transport of *radioactive material* with other dangerous goods, the relevant transport regulations for dangerous goods shall apply in addition to these Regulations.

STRUCTURE

111. Section I presents the background, objective and scope of these Regulations; Section II defines the terms that are required for the purposes of these Regulations; Section III provides general provisions; Section IV provides activity limits and classification used throughout these Regulations; Section V provides requirements and controls for transport; Section VI provides requirements for *radioactive material* and for *packagings* and *packages*; Section VII provides requirements for test procedures; Section VIII provides requirements for *approvals* and administration; the Appendix provides basic radionuclide values for various radionuclides, namely A_1 and A_2 values, activity concentration limits for exempt material, and activity limits for exempt *consignments*; Annex I provides a summary of *approval* and prior notification requirements; Annex II lists relevant prefixes and symbols from the International System of Units; Annex III provides a summary of *conveyances* and *large freight containers* required to be under *exclusive use*; and the Index provides an alphabetical list of topics that are addressed in these Regulations with references to paragraph numbers.

Section II

DEFINITIONS

The following definitions shall apply for the purposes of these Regulations:

A_1 and A_2

201. A_1 shall mean the activity value of *special form radioactive material* that is listed in the Appendix or derived in Section IV and is used to determine the activity limits for the requirements of these Regulations. A_2 shall mean the activity value of *radioactive material*, other than *special form radioactive material*, that is listed in the Appendix or derived in Section IV and is used to determine the activity limits for the requirements of these Regulations.

Aircraft

202. Cargo *aircraft* shall mean any *aircraft*, other than a passenger *aircraft*, that is carrying goods or property.

203. Passenger *aircraft* shall mean an *aircraft* that carries any person other than a crew member, a *carrier's* employee in an official capacity, an authorized representative of an appropriate national authority, or a person accompanying a *consignment* or other cargo.

Approval

204. *Multilateral approval* shall mean *approval* by the relevant *competent authority* of the country of origin of the *design* or *shipment*, as applicable, and also, where the *consignment* is to be transported *through or into* any other country, *approval* by the *competent authority* of that country.

205. *Unilateral approval* shall mean an *approval* of a *design* that is required to be given by the *competent authority* of the country of origin of the *design* only.

Carrier

206. *Carrier* shall mean any person, organization or government undertaking the carriage of *radioactive material* by any means of transport. The term includes both *carriers* for hire or reward (known as common or contract *carriers* in

some countries) and *carriers* on own account (known as private *carriers* in some countries).

Competent authority

207. *Competent authority* shall mean any body or authority designated or otherwise recognized as such for any purpose in connection with these Regulations.

Compliance assurance

208. *Compliance assurance* shall mean a systematic programme of measures applied by a *competent authority* that is aimed at ensuring that the provisions of these Regulations are met in practice.

Confinement system

209. This paragraph has been deleted.

Consignee

210. *Consignee* shall mean any person, organization or government that is entitled to take delivery of a *consignment.*

Consignment

211. *Consignment* shall mean any *package* or *packages*, or load of *radioactive material*, presented by a *consignor* for transport.

Consignor

212. *Consignor* shall mean any person, organization or government that prepares a *consignment* for transport.

Containment system

213. *Containment system* shall mean the assembly of components of the *packaging* specified by the designer as intended to retain the *radioactive material* during transport.

Contamination

214. *Contamination* shall mean the presence of a radioactive substance on a surface in quantities in excess of 0.4 Bq/cm^2 for beta and gamma emitters and *low toxicity alpha emitters*, or 0.04 Bq/cm^2 for all other alpha emitters.

215. *Non-fixed contamination* shall mean *contamination* that can be removed from a surface during routine conditions of transport.

216. *Fixed contamination* shall mean *contamination* other than *non-fixed contamination.*

Conveyance

217. *Conveyance* shall mean:

(a) For transport by road or rail: any *vehicle.*
(b) For transport by water: any *vessel*, or any hold, compartment, or *defined deck area* of a *vessel.*
(c) For transport by air: any *aircraft.*

Criticality safety index

218. *Criticality safety index* (*CSI*) assigned to a *package*, *overpack* or *freight container* containing *fissile material* shall mean a number that is used to provide control over the accumulation of *packages, overpacks* or *freight containers* containing *fissile material.*

Defined deck area

219. *Defined deck area* shall mean the area of the weather deck of a *vessel*, or of a *vehicle* deck of a roll-on/roll-off ship or ferry, that is allocated for the stowage of *radioactive material.*

Design

220. *Design* shall mean the description of *fissile material* excepted under para. 417(f), *special form radioactive material, low dispersible radioactive material, package* or *packaging* that enables such an item to be fully identified. The description may include specifications, engineering drawings,

reports demonstrating compliance with regulatory requirements, and other relevant documentation.

Dose rate

220A. *Dose rate* shall mean the ambient dose equivalent or the directional dose equivalent, as appropriate, per unit time, measured at the point of interest.

Exclusive use

221. *Exclusive use* shall mean the sole use, by a single *consignor*, of a *conveyance* or of a *large freight container*, in respect of which all initial, intermediate and final loading and unloading and *shipment* are carried out in accordance with the directions of the *consignor* or *consignee*, where so required by these Regulations.

Fissile nuclides and *fissile material*

222. *Fissile nuclides* shall mean uranium-233, uranium-235, plutonium-239 and plutonium-241. *Fissile material* shall mean a material containing any of the *fissile nuclides*. Excluded from the definition of *fissile material* are the following:

(a) *Natural uranium* or *depleted uranium* that is unirradiated;
(b) *Natural uranium* or *depleted uranium* that has been irradiated in thermal reactors only;
(c) Material with less than a total of 0.25 g of *fissile nuclides*;
(d) Any combination of (a), (b) and/or (c).

These exclusions are only valid if there is no other material with *fissile nuclides* in the *package* or in the *consignment* if shipped unpackaged.

Freight container — small, large

223. *Freight container* shall mean an article of transport equipment that is:

(a) Of a permanent character and accordingly strong enough to be suitable for repeated use;
(b) Specially designed to facilitate the transport of goods by one or other modes of transport without intermediate reloading;
(c) Designed to be secured and/or readily handled, and having fittings for these purposes.

The term *freight container* does not include the *vehicle*.

A *small freight container* shall mean a *freight container* that has an internal volume of not more than 3 m^3. A *large freight container* shall mean a *freight container* that has an internal volume of more than 3 m^3.

Intermediate bulk container

224. *Intermediate bulk container* (*IBC*) shall mean a portable *packaging* that:

(a) Has a capacity of not more than 3 m^3;
(b) Is designed for mechanical handling;
(c) Is resistant to the stresses produced in handling and transport, as determined by tests.

Low dispersible radioactive material

225. *Low dispersible radioactive material* shall mean either a solid *radioactive material* or a solid *radioactive material* in a sealed capsule that has limited dispersibility and is not in powder form.

Low specific activity material

226. *Low specific activity* (*LSA*) *material* shall mean *radioactive material* that by its nature has a limited *specific activity*, or *radioactive material* for which limits of estimated average *specific activity* apply. External shielding materials surrounding the *LSA material* shall not be considered in determining the estimated average *specific activity*.

Low toxicity alpha emitters

227. *Low toxicity alpha emitters* are: *unirradiated uranium* enriched up to 20%, *natural uranium, depleted uranium*, natural thorium, uranium-235, uranium-238 and thorium-232; or alpha emitters with a half-life of less than 10 days. Thorium-228 and thorium-230 may be included when contained in ores or in physical and chemical concentrates.

Management system

228. *Management system* shall mean a set of interrelated or interacting elements (system) for establishing policies and objectives and enabling the objectives to be achieved in an efficient and effective manner.

Maximum normal operating pressure

229. *Maximum normal operating pressure* shall mean the maximum pressure above atmospheric pressure at mean sea level that would develop in the *containment system* in a period of one year under the conditions of temperature and solar radiation corresponding to the environmental conditions in the absence of venting, external cooling by an ancillary system, or operational controls during transport.

Naturally occurring radioactive material (*NORM*)

229A. *Naturally occurring radioactive material* (*NORM*) shall mean *radioactive material* containing no significant amounts of radionuclides other than *naturally occurring radionuclides*. The amounts of radionuclides other than *naturally occurring radionuclides* shall not exceed the exemption values specified in the Appendix, or as calculated in accordance with paras 403–407. Material in which the activity concentrations of the *naturally occurring radionuclides* have been changed by a process is included.

Naturally occurring radionuclides

229B. *Naturally occurring radionuclides* shall mean radionuclides that occur naturally on Earth in significant quantities. The term is used to refer to the primordial radionuclides potassium-40, uranium-235, uranium-238 and thorium-232 and their radioactive decay products. This includes radionuclides found in U (natural) and Th (natural).

Overpack

230. *Overpack* shall mean an enclosure used by a single *consignor* to contain one or more *packages,* and to form one unit for convenience of handling and stowage during transport.

Package

231. *Package* shall mean the complete product of the packing operation, consisting of the *packaging* and its contents prepared for transport. The types of *packages* covered by these Regulations that are subject to the activity limits and classification in Section IV and meet the corresponding requirements are:

(a) *Excepted package*;
(b) *Industrial package Type 1 (Type IP-1)*;
(c) *Industrial package Type 2 (Type IP-2)*;
(d) *Industrial package Type 3 (Type IP-3)*;
(e) *Type A package*;
(f) *Type B(U) package*;
(g) *Type B(M) package*;
(h) *Type C package.*

Packages containing *fissile material* or uranium hexafluoride are subject to additional requirements.

Packaging

232. *Packaging* shall mean one or more receptacles and any other components or materials necessary for the receptacles to perform containment and other safety functions.

Radiation level

233. This paragraph was deleted and its content was transferred to para. 220A.

Radiation protection programme

234. *Radiation protection programme* shall mean systematic arrangements that are aimed at providing adequate consideration of radiation protection measures.

Radioactive contents

235. *Radioactive contents* shall mean the *radioactive material* together with any contaminated or activated solids, liquids and gases within the *packaging*.

Radioactive material

236. *Radioactive material* shall mean any material containing radionuclides where both the activity concentration and the total activity in the *consignment* exceed the values specified in paras 402–407.

Shipment

237. *Shipment* shall mean the specific movement of a *consignment* from origin to destination.

Special arrangement

238. *Special arrangement* shall mean those provisions, approved by the *competent authority*, under which *consignments* that do not satisfy all the applicable requirements of these Regulations may be transported.

Special form radioactive material

239. *Special form radioactive material* shall mean either an indispersible solid *radioactive material* or a sealed capsule containing *radioactive material*.

Specific activity

240. *Specific activity* of a radionuclide shall mean the activity per unit mass of that nuclide. The *specific activity* of a material shall mean the activity per unit mass of the material in which the radionuclides are essentially uniformly distributed.

Note: The terms 'activity concentration' and *specific activity* of a material are synonymous for the purpose of these Regulations.

Surface contaminated object

241. *Surface contaminated object* (*SCO*) shall mean a solid object that is not itself radioactive but which has *radioactive material* distributed on its surface.

Tank

242. *Tank* shall mean a portable *tank* (including a *tank* container), a road *tank vehicle*, a rail *tank* wagon or a receptacle that contains solids, liquids or gases, having a capacity of not less than 450 L when used for the transport of gases.

Through or into

243. *Through or into* shall mean *through or into* the countries in which a *consignment* is transported but specifically excludes countries over which a *consignment* is carried by air, provided that there are no scheduled stops in those countries.

Transport index

244. *Transport index* (*TI*) assigned to a *package*, *overpack* or *freight container*, or to unpackaged *LSA-I*, *SCO-I* or *SCO-III*, shall mean a number that is used to provide control over radiation exposure.

Unirradiated thorium

245. *Unirradiated thorium* shall mean thorium containing not more than 10^{-7} g of uranium-233 per gram of thorium-232.

Unirradiated uranium

246. *Unirradiated uranium* shall mean uranium containing not more than 2×10^3 Bq of plutonium per gram of uranium-235, not more than 9×10^6 Bq of fission products per gram of uranium-235 and not more than 5×10^{-3} g of uranium-236 per gram of uranium-235.

Uranium — natural, depleted, enriched

247. *Natural uranium* shall mean uranium (which may be chemically separated) containing the naturally occurring distribution of uranium isotopes (approximately 99.28% uranium-238 and 0.72% uranium-235, by mass).

Depleted uranium shall mean uranium containing a lesser mass percentage of uranium-235 than in *natural uranium*.

Enriched uranium shall mean uranium containing a greater mass percentage of uranium-235 than 0.72%.

In all cases, a very small mass percentage of uranium-234 is present.

Vehicle

248. *Vehicle* shall mean a road *vehicle* (including an articulated *vehicle,* i.e. a tractor and semi-trailer combination), railroad car or railway wagon. Each trailer shall be considered as a separate *vehicle.*

Vessel

249. *Vessel* shall mean any sea-going *vessel* or inland waterway craft used for carrying cargo.

Section III

GENERAL PROVISIONS

RADIATION PROTECTION

301. Doses to persons shall be below the relevant dose limits. Protection and safety shall be optimized in order that the magnitude of individual doses, the number of persons exposed and the likelihood of incurring exposure shall be kept as low as reasonably achievable, economic and social factors being taken into account, within the restriction that the doses to individuals are subject to dose constraints in accordance with GSR Part 3 [2]. A structured and systematic approach shall be adopted and shall include consideration of the interfaces between transport and other activities.

302. A *radiation protection programme* shall be established for the transport of *radioactive material*. The nature and extent of the measures to be employed in the programme shall be related to the magnitude and likelihood of radiation exposure. The programme shall incorporate the requirements of paras 301, 303–305, 311 and 562. Programme documents shall be available, on request, for inspection by the relevant *competent authority*.

303. For occupational exposures arising from transport activities, where it is assessed that the effective dose either:

(a) Is likely to be between 1 and 6 mSv in a year, a dose assessment programme via workplace monitoring or individual monitoring shall be conducted; or
(b) Is likely to exceed 6 mSv in a year, individual monitoring shall be conducted.

When workplace monitoring or individual monitoring is conducted, appropriate records shall be kept.

EMERGENCY RESPONSE

304. In the event of a nuclear or radiological emergency during the transport of *radioactive material*, provisions as established by relevant national and/or international organizations shall be observed to protect people, property and the environment. These provisions include arrangements for preparedness and response established in accordance with the national and/or international

requirements and in a consistent and coordinated manner with the national and/or international emergency arrangements.

305. The arrangements for preparedness and response shall be based on the graded approach and shall take into consideration the identified hazards and their potential consequences, including the formation of other dangerous substances that may result from the reaction between the contents of a *consignment* and the environment in the event of a nuclear or radiological emergency. Requirements for the establishment of such arrangements are provided in GSR Part 7 [6], and supporting recommendations are provided in SSG-65 [8] and IAEA Safety Standards Series Nos GSG-2, Criteria for Use in Preparedness and Response for a Nuclear or Radiological Emergency [14]; GS-G-2.1, Arrangements for Preparedness for a Nuclear or Radiological Emergency [15]; and GSG-11, Arrangements for the Termination of a Nuclear or Radiological Emergency [16].

MANAGEMENT SYSTEM

306. A *management system* based on international, national or other standards acceptable to the *competent authority* shall be established and implemented for all activities within the scope of these Regulations, as identified in para. 106, to ensure compliance with the relevant provisions of these Regulations. Certification that the *design* specification has been fully implemented shall be available to the *competent authority*. The manufacturer, *consignor* or user shall be prepared:

(a) To provide facilities for inspection during manufacture and use;
(b) To demonstrate compliance with these Regulations to the *competent authority*.

Where *competent authority approval* is required, such *approval* shall take into account, and be contingent upon, the adequacy of the *management system*.

COMPLIANCE ASSURANCE

307. The *competent authority* shall assure compliance with these Regulations.

308. The relevant *competent authority* shall arrange for periodic assessments of the radiation doses to persons due to the transport of *radioactive material*, to ensure that the system of protection and safety complies with GSR Part 3 [2].

NON-COMPLIANCE

309. In the event of non-compliance with any limit in these Regulations applicable to *dose rate* or *contamination*:

(a) The *consignor*, *consignee*, *carrier* and any organization involved during transport that may be affected, as appropriate, shall be informed of the non-compliance by:
   (i) The *carrier* if the non-compliance is identified during transport; or
   (ii) The *consignee* if the non-compliance is identified at receipt.
(b) The *consignor*, *carrier*, or *consignee*, as appropriate, shall:
   (i) Take immediate steps to mitigate the consequences of the non-compliance;
   (ii) Investigate the non-compliance and its causes, circumstances and consequences;
   (iii) Take appropriate action to remedy the causes and circumstances that led to the non-compliance and to prevent a recurrence of the causes and circumstances similar to those that led to the non-compliance;
   (iv) Communicate to the relevant *competent authority* the causes of the non-compliance and the corrective or preventive actions taken or to be taken.
(c) The communication of the non-compliance to the *consignor* and the relevant *competent authority*, respectively, shall be made as soon as practicable and shall be immediate whenever an emergency exposure situation has developed or is developing.

SPECIAL ARRANGEMENT

310. *Consignments* for which conformity with the other provisions of these Regulations is impracticable shall not be transported except under *special arrangement*. Provided the *competent authority* is satisfied that conformity with the other provisions of these Regulations is impracticable and that the requisite standards of safety established by these Regulations have been demonstrated through means alternative to the other provisions of these Regulations, the *competent authority* may approve *special arrangement* transport operations for a single *consignment* or a planned series of multiple *consignments*. The overall level of safety in transport shall be at least equivalent to that which would be provided if all the applicable requirements in these Regulations had been met. For *consignments* of this type, *multilateral approval* shall be required.

TRAINING

311. Workers shall receive appropriate training concerning radiation protection, including the precautions to be observed in order to restrict their occupational exposure and the exposure of other persons who might be affected by their actions.

312. Persons engaged in the transport of *radioactive material* shall receive training on the contents of these Regulations commensurate with their responsibilities.

313. Persons such as those who classify *radioactive material*; pack *radioactive material*; mark and label *radioactive material*; prepare transport documents for *radioactive material*; offer or accept *radioactive material* for transport; carry or handle *radioactive material* during transport; mark or placard or load or unload *packages* of *radioactive material* into or from transport *vehicles*, bulk *packagings* or *freight containers*; or are otherwise directly involved in the transport of *radioactive material* as determined by the *competent authority* shall receive the following training:

(a) General awareness/familiarization training:
   - (i) Each person shall receive training designed to provide familiarity with the general provisions of these Regulations.
   - (ii) The general awareness/familiarization training shall include a description of the classification of *radioactive material* in accordance with Section IV; labelling, marking, placarding and *packaging* and segregation requirements; the purpose and content of the *radioactive material* transport document; and the available emergency response documents.

(b) Function specific training: Each person shall receive detailed training concerning specific *radioactive material* transport requirements that are applicable to the function that person performs.

(c) Safety training: Commensurate with the risk of exposure in the event of a release, and with the functions performed, each person shall receive training on:
   - (i) Methods and procedures for avoidance of accident conditions during transport, such as proper use of *package* handling equipment and appropriate methods of stowage of *radioactive material*.
   - (ii) Available emergency response information and how to use it.
   - (iii) General hazards presented by the various classifications of *radioactive material* in accordance with Section IV and how to prevent exposure to those hazards, including, if appropriate, the use of personal protective clothing and equipment.

(iv) Procedures to be immediately followed in the event of an unintentional release of *radioactive material*, including any emergency response procedures for which the person is responsible and personal protection procedures to be followed.

314. Records of all safety training undertaken shall be kept by the employer and made available to the employee if requested.

315. The training required in para. 313 shall be provided or verified upon employment in a position involving *radioactive material* transport and shall be periodically supplemented with retraining.

Section IV

ACTIVITY LIMITS AND CLASSIFICATION

GENERAL PROVISIONS

401. *Radioactive material* shall be assigned one of the United Nations (UN) numbers specified in Table 1 in accordance with paras 408–434.

TABLE 1. EXCERPTS FROM THE LIST OF UN NUMBERS, PROPER SHIPPING NAMES AND DESCRIPTIONS

Assignment of UN numbers	PROPER SHIPPING NAME and description[a]
Excepted packages	
UN 2908	RADIOACTIVE MATERIAL, EXCEPTED PACKAGE — EMPTY PACKAGING
UN 2909	RADIOACTIVE MATERIAL, EXCEPTED PACKAGE — ARTICLES MANUFACTURED FROM NATURAL URANIUM or DEPLETED URANIUM or NATURAL THORIUM
UN 2910	RADIOACTIVE MATERIAL, EXCEPTED PACKAGE — LIMITED QUANTITY OF MATERIAL
UN 2911	RADIOACTIVE MATERIAL, EXCEPTED PACKAGE — INSTRUMENTS or ARTICLES
UN 3507	URANIUM HEXAFLUORIDE, RADIOACTIVE MATERIAL, EXCEPTED PACKAGE, less than 0.1 kg per package, non-fissile or fissile-excepted[b]
Low specific activity material	
UN 2912	RADIOACTIVE MATERIAL, LOW SPECIFIC ACTIVITY (LSA-I), non-fissile or fissile-excepted[b]
UN 3321	RADIOACTIVE MATERIAL, LOW SPECIFIC ACTIVITY (LSA-II), non-fissile or fissile-excepted[b]

TABLE 1. EXCERPTS FROM THE LIST OF UN NUMBERS, PROPER SHIPPING NAMES AND DESCRIPTIONS (cont.)

Assignment of UN numbers	PROPER SHIPPING NAME and description[a]
UN 3322	RADIOACTIVE MATERIAL, LOW SPECIFIC ACTIVITY (LSA-III), non-fissile or fissile-excepted[b]
UN 3324	RADIOACTIVE MATERIAL, LOW SPECIFIC ACTIVITY (LSA-II), FISSILE
UN 3325	RADIOACTIVE MATERIAL, LOW SPECIFIC ACTIVITY (LSA-III), FISSILE
Surface contaminated objects	
UN 2913	RADIOACTIVE MATERIAL, SURFACE CONTAMINATED OBJECTS (SCO-I, SCO-II or SCO-III), non-fissile or fissile-excepted[b]
UN 3326	RADIOACTIVE MATERIAL, SURFACE CONTAMINATED OBJECTS (SCO-I or SCO-II), FISSILE
Type A packages	
UN 2915	RADIOACTIVE MATERIAL, TYPE A PACKAGE, non-special form, non-fissile or fissile-excepted[b]
UN 3327	RADIOACTIVE MATERIAL, TYPE A PACKAGE, FISSILE, non-special form
UN 3332	RADIOACTIVE MATERIAL, TYPE A PACKAGE, SPECIAL FORM, non-fissile or fissile-excepted[b]
UN 3333	RADIOACTIVE MATERIAL, TYPE A PACKAGE, SPECIAL FORM, FISSILE
Type B(U) packages	
UN 2916	RADIOACTIVE MATERIAL, TYPE B(U) PACKAGE, non-fissile or fissile-excepted[b]
UN 3328	RADIOACTIVE MATERIAL, TYPE B(U) PACKAGE, FISSILE

TABLE 1. EXCERPTS FROM THE LIST OF UN NUMBERS, PROPER SHIPPING NAMES AND DESCRIPTIONS (cont.)

Assignment of UN numbers	PROPER SHIPPING NAME and description[a]
Type B(M) packages	
UN 2917	RADIOACTIVE MATERIAL, TYPE B(M) PACKAGE, non-fissile or fissile-excepted[b]
UN 3329	RADIOACTIVE MATERIAL, TYPE B(M) PACKAGE, FISSILE
Type C packages	
UN 3323	RADIOACTIVE MATERIAL, TYPE C PACKAGE, non-fissile or fissile-excepted[b]
UN 3330	RADIOACTIVE MATERIAL, TYPE C PACKAGE, FISSILE
Special arrangement	
UN 2919	RADIOACTIVE MATERIAL, TRANSPORTED UNDER SPECIAL ARRANGEMENT, non-fissile or fissile-excepted[b]
UN 3331	RADIOACTIVE MATERIAL, TRANSPORTED UNDER SPECIAL ARRANGEMENT, FISSILE
Uranium hexafluoride	
UN 2977	RADIOACTIVE MATERIAL, URANIUM HEXAFLUORIDE, FISSILE
UN 2978	RADIOACTIVE MATERIAL, URANIUM HEXAFLUORIDE, non-fissile or fissile-excepted[b]
UN 3507	URANIUM HEXAFLUORIDE, RADIOACTIVE MATERIAL, EXCEPTED PACKAGE, less than 0.1 kg per package, non-fissile or fissile-excepted[b]

[a] The “PROPER SHIPPING NAME” is found in the column “PROPER SHIPPING NAME and description” and is restricted to that part shown in CAPITAL LETTERS. In the cases of UN 2909, UN 2911, UN 2913 and UN 3326, where alternative proper shipping names are separated by the word “or”, only the relevant proper shipping name shall be used.

[b] The term ‘fissile-excepted’ refers only to material excepted under para. 417.

BASIC RADIONUCLIDE VALUES

402. The following basic values for individual radionuclides are given in the Appendix:

(a) A_1 and A_2 in TBq;
(b) Activity concentration limits for exempt material in Bq/g;
(c) Activity limits for exempt *consignments* in Bq.

DETERMINATION OF BASIC RADIONUCLIDE VALUES

403. For individual radionuclides:

(a) That are not listed in the Appendix, the determination of the basic radionuclide values referred to in para. 402 shall require *multilateral approval*. For these radionuclides, activity concentrations for exempt material and activity limits for exempt *consignments* shall be calculated in accordance with the principles established in GSR Part 3 [2]. It is permissible to use an A_2 value calculated using a dose coefficient for the appropriate lung absorption type, as recommended by the International Commission on Radiological Protection, if the chemical forms of each radionuclide under both normal and accident conditions of transport are taken into consideration. Alternatively, the radionuclide values in Table 3 may be used without obtaining *competent authority approval*.
(b) In instruments or articles in which the *radioactive material* is enclosed in or is included as a component part of the instrument or other manufactured article and which meets para. 423(c), alternative basic radionuclide values to those in the Appendix for the activity limit for an exempt *consignment* are permitted and shall require *multilateral approval*. Such alternative activity limits for an exempt *consignment* shall be calculated in accordance with the principles set out in GSR Part 3 [2].

404. In the calculations of A_1 and A_2 for a radionuclide not listed in the Appendix, a single radioactive decay chain in which the radionuclides are present in their naturally occurring proportions, and in which no progeny nuclide has a half-life either longer than 10 days or longer than that of the parent nuclide, shall be considered as a single radionuclide; the activity to be taken into account and the A_1 or A_2 value to be applied shall be that corresponding to the parent nuclide of that chain. In the case of radioactive decay chains in which any progeny nuclide has a half-life either longer than 10 days or longer than that of the parent

nuclide, the parent and such progeny nuclides shall be considered as mixtures of different nuclides.

405. For mixtures of radionuclides, the basic radionuclide values referred to in para. 402 may be determined as follows:

$$X_m = \frac{1}{\sum_i \frac{f(i)}{X(i)}}$$

where

$f(i)$ is the fraction of activity or activity concentration of radionuclide i in the mixture;

$X(i)$ is the appropriate value of A_1 or A_2, or the activity concentration limit for exempt material or the activity limit for an exempt *consignment* as appropriate for radionuclide i;

X_m is the derived value of A_1 or A_2, or the activity concentration limit for exempt material or the activity limit for an exempt *consignment* in the case of a mixture.

TABLE 2. This table has been deleted and its content has been incorporated in the Appendix.

406. When the identity of each radionuclide is known but the individual activities of some of the radionuclides are not known, the radionuclides may be grouped and the lowest radionuclide value, as appropriate for the radionuclides in each group, may be used in applying the formulas in paras 405 and 430. Groups may be based on the total alpha activity and the total beta/gamma activity, when these are known, using the lowest radionuclide values for the alpha emitters or beta/gamma emitters, respectively.

407. For individual radionuclides or for mixtures of radionuclides for which relevant data are not available, the values shown in Table 3 shall be used.

TABLE 3. BASIC RADIONUCLIDE VALUES FOR UNKNOWN RADIONUCLIDES OR MIXTURES

Radioactive contents	A_1 (TBq)	A_2 (TBq)	Activity concentration limit for exempt material (Bq/g)	Activity limit for an exempt *consignment* (Bq)
Only beta or gamma emitting nuclides are known to be present	1×10^{-1}	1×10^{-3}	1×10^{1}	1×10^{4}
Alpha emitting nuclides, but no neutron emitters are known to be present	2×10^{-1}	3×10^{-4}	1×10^{-1}	1×10^{3}
Neutron emitting nuclides are known to be present or no relevant data are available	4×10^{-3}	8×10^{-5}	1×10^{-1}	1×10^{3}

CLASSIFICATION OF MATERIAL

Low specific activity material

408. *Radioactive material* may only be classified as *LSA material* if the conditions of paras 226, 409–411 and 517–522 are met.

409. *LSA material* shall be in one of three groups:

(a) *LSA-I*:
   - (i) Uranium and thorium ores and concentrates of such ores, and other ores containing *naturally occurring radionuclides*.
   - (ii) *Natural uranium*, *depleted uranium*, natural thorium or their compounds or mixtures, that are unirradiated and in solid or liquid form.
   - (iii) *Radioactive material* for which the A_2 value is unlimited. *Fissile material* may be included only if excepted under para. 417.
   - (iv) Other *radioactive material* in which the activity is distributed throughout and the estimated average *specific activity* does not

exceed 30 times the values for the activity concentration specified in paras 402–407. *Fissile material* may be included only if excepted under para. 417.

(b) *LSA-II*:
   (i) Tritiated water with an activity concentration of up to 0.8 TBq/L;
   (ii) Other material in which the activity is distributed throughout and the estimated average *specific activity* does not exceed $10^{-4}A_2$/g for solids and gases, and $10^{-5}A_2$/g for liquids.

(c) *LSA-III*:
   Solids (e.g. consolidated wastes, activated materials), excluding powders, in which:
   (i) The *radioactive material* is distributed throughout a solid or a collection of solid objects or is essentially uniformly distributed in a solid compact binding agent (e.g. concrete, bitumen, ceramic).
   (ii) The estimated average *specific activity* of the solid, excluding any shielding material, does not exceed $2 \times 10^{-3}A_2$/g.

410. A single *package* of non-combustible solid *LSA-II* or *LSA-III material*, if carried by air, shall not contain an activity greater than $3000A_2$.

411. The *radioactive contents* in a single *package* of *LSA material* shall be so restricted that the *dose rate* specified in para. 517 shall not be exceeded, and the activity in a single *package* shall also be so restricted that the activity limits for a *conveyance* specified in para. 522 shall not be exceeded.

Surface contaminated object

412. *Radioactive material* may be classified as *SCO* if the conditions in paras 241, 413, 414 and 517–522 are met.

413. *SCO* shall be in one of three groups:

(a) *SCO-I*: A solid object on which:
   (i) The *non-fixed contamination* on the accessible surface averaged over 300 cm^2 (or the area of the surface if less than 300 cm^2) does not exceed 4 Bq/cm^2 for beta and gamma emitters and *low toxicity alpha emitters*, or 0.4 Bq/cm^2 for all other alpha emitters.
   (ii) The *fixed contamination* on the accessible surface averaged over 300 cm^2 (or the area of the surface if less than 300 cm^2) does not exceed 4×10^4 Bq/cm^2 for beta and gamma emitters and *low toxicity alpha emitters*, or 4000 Bq/cm^2 for all other alpha emitters.

(iii) The *non-fixed contamination* plus the *fixed contamination* on the inaccessible surface averaged over 300 cm^2 (or the area of the surface if less than 300 cm^2) does not exceed 4×10^4 Bq/cm^2 for beta and gamma emitters and *low toxicity alpha emitters*, or 4000 Bq/cm^2 for all other alpha emitters.

(b) *SCO-II*: A solid object on which either the *fixed* or *non-fixed contamination* on the surface exceeds the applicable limits specified for *SCO-I* in subparagraph (a) above and on which:

(i) The *non-fixed contamination* on the accessible surface averaged over 300 cm^2 (or the area of the surface if less than 300 cm^2) does not exceed 400 Bq/cm^2 for beta and gamma emitters and *low toxicity alpha emitters*, or 40 Bq/cm^2 for all other alpha emitters.

(ii) The *fixed contamination* on the accessible surface averaged over 300 cm^2 (or the area of the surface if less than 300 cm^2) does not exceed 8×10^5 Bq/cm^2 for beta and gamma emitters and *low toxicity alpha emitters*, or 8×10^4 Bq/cm^2 for all other alpha emitters.

(iii) The *non-fixed contamination* plus the *fixed contamination* on the inaccessible surface averaged over 300 cm^2 (or the area of the surface if less than 300 cm^2) does not exceed 8×10^5 Bq/cm^2 for beta and gamma emitters and *low toxicity alpha emitters*, or 8×10^4 Bq/cm^2 for all other alpha emitters.

(c) *SCO-III*: A large solid object that, because of its size, cannot be transported in a type of *package* described in these Regulations and for which:

(i) All openings are sealed to prevent release of *radioactive material* during the conditions specified in para. 520(e).

(ii) The inside of the object is as dry as practicable.

(iii) The *non-fixed contamination* on the external surfaces does not exceed the limits specified in para. 508.

(iv) The *non-fixed contamination* plus the *fixed contamination* on the inaccessible surface averaged over 300 cm^2 does not exceed 8×10^5 Bq/cm^2 for beta and gamma emitters and *low toxicity alpha emitters*, or 8×10^4 Bq/cm^2 for all other alpha emitters.

414. The *radioactive contents* in a single *package* of *SCO* shall be so restricted that the *dose rate* specified in para. 517 shall not be exceeded, and the activity in a single *package* shall also be so restricted that the activity limits for a *conveyance* specified in para. 522 shall not be exceeded.

Special form radioactive material

415. *Radioactive material* may be classified as *special form radioactive material* only if it meets the definition in para. 239 and the requirements of paras 602–604A and 802.

Low dispersible radioactive material

416. *Radioactive material* may be classified as *low dispersible radioactive material* only if it meets the definition in para. 225 and the requirements of para. 605, taking into account the requirements of paras 665 and 802.

Fissile material

417. *Fissile material* and *packages* containing *fissile material* shall be classified under the relevant entry as "FISSILE" in accordance with Table 1 unless excepted by one of the provisions of subparagraphs (a)–(h) of this paragraph and transported subject to the requirements of para. 570. All provisions apply only to material in *packages* that meet the requirements of para. 636, unless unpackaged material is specifically allowed in the provision. The provisions for the exceptions are as follows:

(a) Uranium enriched in uranium-235 to a maximum of 1% by mass, and with a total plutonium and uranium-233 content not exceeding 1% of the mass of uranium-235, provided that the *fissile nuclides* are distributed essentially homogeneously throughout the material. In addition, if uranium-235 is present in metallic, oxide or carbide forms, it shall not form a lattice arrangement.

(b) Liquid solutions of uranyl nitrate enriched in uranium-235 to a maximum of 2% by mass, with a total plutonium and uranium-233 content not exceeding 0.002% of the mass of uranium, and with a minimum nitrogen to uranium atomic ratio (N/U) of 2.

(c) Uranium with a maximum uranium enrichment of 5% by mass of uranium-235 provided that:
    (i) There is no more than 3.5 g of uranium-235 per *package*.
    (ii) The total plutonium and uranium-233 content does not exceed 1% of the mass of uranium-235 per *package*.
    (iii) Transport of the *package* is subject to the *consignment* limit provided in para. 570(c).

(d) *Fissile nuclides* with a total mass not greater than 2.0 g per *package*, provided that the *package* is transported subject to the *consignment* limit provided in para. 570(d).
(e) *Fissile nuclides* with a total mass not greater than 45 g, either packaged or unpackaged, subject to the requirements of para. 570(e).
(f) A *fissile material* that meets the requirements of paras 570(b), 606 and 802.
(g) *Enriched uranium* in the form of residual *contamination* on inner surfaces of clean and washed out cylinders that conform to the International Organization for Standardization document: Nuclear Energy — Packagings for the Transport of Uranium Hexafluoride (UF_6) (ISO 7195) [17] and that have contained enriched uranium hexafluoride, provided that the average amount of uranium-235 on the inner surfaces does not exceed 2.5 g/m^2, with a total mass of *fissile nuclides* not exceeding 15 g per *package*.
(h) *Fissile nuclides* with a total mass not greater than 15 g per *package*, provided that:
    (i) The mass of *fissile nuclides* does not exceed 0.5% of the mass of solid material of the *package* (including *packaging* material).
    (ii) The *package* is transported subject to the requirements of para. 570(f).
    (iii) Lead, beryllium, hydrogenous material enriched in deuterium, graphite and other allotropic forms of carbon that may be present in the *package* shall not be included in determining the required mass of solid material.

418. The contents of *packages* containing *fissile material* shall be as specified for the *package design*, either directly in these Regulations or in the certificate of *approval*.

Uranium hexafluoride

419. Uranium hexafluoride shall be assigned to one of the following UN numbers only:

(a) UN 2977, RADIOACTIVE MATERIAL, URANIUM HEXAFLUORIDE, FISSILE;
(b) UN 2978, RADIOACTIVE MATERIAL, URANIUM HEXAFLUORIDE, non-fissile or fissile-excepted;
(c) UN 3507, URANIUM HEXAFLUORIDE, RADIOACTIVE MATERIAL, EXCEPTED PACKAGE, less than 0.1 kg per package, non-fissile or fissile-excepted.

420. The contents of a *package* containing uranium hexafluoride shall comply with the following requirements:

(a) The mass of uranium hexafluoride shall not be different from that allowed for by the *package design*.
(b) The mass of uranium hexafluoride shall not be greater than a value that would lead to an ullage of less than 5% at the maximum temperature of the *package*, as specified for in the plant systems where the *package* might be used.
(c) The uranium hexafluoride shall be in solid form and the internal pressure shall not be above atmospheric pressure when presented for transport.

CLASSIFICATION OF PACKAGES

421. The quantity of *radioactive material* in a *package* shall not exceed the relevant limits for the *package* type as specified below.

Classification as excepted package

422. A *package* may only be classified as an *excepted package* if it meets the requirements of para. 516 and one of the following conditions:

(a) It is an empty *packaging* that had previously contained *radioactive material* and meets the requirements of para. 427.
(b) It contains instruments or articles and meets the requirements of para. 423.
(c) It contains articles manufactured from *natural uranium*, *depleted uranium* or natural thorium and meets the requirements of para. 426.
(d) It contains *radioactive material* and meets the requirements of para. 424.
(e) It contains less than 0.1 kg of uranium hexafluoride and meets the requirements of para. 425.

423. *Radioactive material* that is enclosed in or is included as a component part of an instrument or other manufactured article may be classified under UN 2911, RADIOACTIVE MATERIAL, EXCEPTED PACKAGE — INSTRUMENTS or ARTICLES, provided that:

(a) The *dose rate* at 10 cm from any point on the external surface of any unpackaged instrument or article is not greater than 0.1 mSv/h.

(b) Each instrument or article bears the mark "RADIOACTIVE" on its external surface except for the following:
   (i) Radioluminescent timepieces or devices do not require marks.
   (ii) Consumer products that either have received regulatory authorization in accordance with para. 107(e) or do not individually exceed the activity limit for an exempt *consignment* specified in column 5 of the Appendix do not require marks, provided that such products are transported in a *package* that bears the mark "RADIOACTIVE" on its internal surface in such a manner that a warning of the presence of *radioactive material* is visible on opening the *package*.
   (iii) Other instruments or articles too small to bear the mark "RADIOACTIVE" do not require marks, provided that they are transported in a *package* that bears the mark "RADIOACTIVE" on its internal surface in such a manner that a warning of the presence of *radioactive material* is visible on opening the *package*.

(c) The active material is completely enclosed by non-active components. (A device performing the sole function of containing *radioactive material* shall not be considered to be an instrument or manufactured article.)
(d) The limits specified in columns 2 and 3 of Table 4 are met for each individual item and each *package*, respectively.
(e) For transport by post, the total activity in each *excepted package* shall not exceed one tenth of the relevant limits specified in column 3 of Table 4.
(f) If the *package* contains *fissile material*, one of the provisions of subparagraphs (a)–(h) of para. 417 shall apply.

424. *Radioactive material* in forms other than as specified in para. 423 and with an activity not exceeding the limits specified in column 4 of Table 4 may be classified under UN 2910, RADIOACTIVE MATERIAL, EXCEPTED PACKAGE — LIMITED QUANTITY OF MATERIAL, provided that:

(a) The *package* retains its *radioactive contents* under routine conditions of transport.
(b) The *package* bears the mark "RADIOACTIVE" on either:
   (i) An internal surface in such a manner that a warning of the presence of *radioactive material* is visible on opening the *package*; or
   (ii) The outside of the *package*, where it is impractical to mark an internal surface.

(c) For transport by post, the total activity in each *excepted package* shall not exceed one tenth of the relevant limits specified in column 4 of Table 4.
(d) If the *package* contains *fissile material*, one of the provisions of subparagraphs (a)–(h) of para. 417 shall apply.

TABLE 4. ACTIVITY LIMITS FOR EXCEPTED PACKAGES

Physical state of contents	Instrument or article		Materials
	Item limits[a]	*Package* limits[a]	*Package* limits[a]
Solids			
Special form	$10^{-2}A_1$	A_1	$10^{-3}A_1$
Other forms	$10^{-2}A_2$	A_2	$10^{-3}A_2$
Liquids	$10^{-3}A_2$	$10^{-1}A_2$	$10^{-4}A_2$
Gases			
Tritium	$2 \times 10^{-2}A_2$	$2 \times 10^{-1}A_2$	$2 \times 10^{-2}A_2$
Special form	$10^{-3}A_1$	$10^{-2}A_1$	$10^{-3}A_1$
Other forms	$10^{-3}A_2$	$10^{-2}A_2$	$10^{-3}A_2$

[a] For mixtures of radionuclides, see paras 405–407.

425. Uranium hexafluoride not exceeding the limits specified in column 4 of Table 4 may be classified under UN 3507, URANIUM HEXAFLUORIDE, RADIOACTIVE MATERIAL, EXCEPTED PACKAGE, less than 0.1 kg per package, non-fissile or fissile-excepted, provided that:

(a) The mass of uranium hexafluoride in the *package* is less than 0.1 kg.
(b) The conditions of paras 420, 424(a) and 424(b) are met.

426. Articles manufactured from *natural uranium*, *depleted uranium* or natural thorium and articles in which the sole *radioactive material* is unirradiated *natural uranium*, unirradiated *depleted uranium* or unirradiated natural thorium may be classified under UN 2909, RADIOACTIVE MATERIAL, EXCEPTED PACKAGE — ARTICLES MANUFACTURED FROM NATURAL URANIUM or DEPLETED URANIUM or NATURAL THORIUM, provided that the outer surface of the uranium or thorium is enclosed in an inactive sheath made of metal or some other substantial material.

Additional requirements and controls for transport of empty packagings

427. An empty *packaging* that had previously contained *radioactive material* may be classified under UN 2908, RADIOACTIVE MATERIAL, EXCEPTED PACKAGE — EMPTY PACKAGING, provided that:

(a) It is in a well-maintained condition and securely closed.
(b) The outer surface of any uranium or thorium in its structure is covered with an inactive sheath made of metal or some other substantial material.
(c) The level of internal *non-fixed contamination* does not exceed 100 times the levels specified in para. 508.
(d) Any labels that may have been displayed on it in conformity with para. 538 are no longer visible.
(e) If the *packaging* has contained *fissile material*, one of the provisions of subparagraphs (a)–(h) of para. 417 or one of the provisions for exclusion in para. 222 shall apply.

Classification as Type A package

428. *Packages* containing *radioactive material* may be classified as *Type A packages* provided that the conditions of paras 429 and 430 are met.

429. *Type A packages* shall not contain activities greater than either of the following:

(a) For *special form radioactive material* — A_1;
(b) For all other *radioactive material* — A_2.

430. For mixtures of radionuclides whose identities and respective activities are known, the following condition shall apply to the *radioactive contents* of a *Type A package*:

$$\sum_{i} \frac{B(i)}{A_1(i)} + \sum_{j} \frac{C(j)}{A_2(j)} \leq 1$$

where

$B(i)$ is the activity of radionuclide i as *special form radioactive material*;
$A_1(i)$ is the A_1 value for radionuclide i;

$C(\mathrm{j})$ is the activity of radionuclide j as other than *special form radioactive material*;

$A_2(\mathrm{j})$ is the A_2 value for radionuclide j.

Classification as Type B(U), Type B(M) or Type C package

431. *Type B(U), Type B(M)* and *Type C packages* shall be classified in accordance with the *competent authority* certificate of *approval* for the *package design* issued by the country of origin of the *design*.

432. The contents of a *Type B(U), Type B(M)* or *Type C package* shall be as specified in the certificate of *approval*.

433. *Type B(U)* and *Type B(M) packages*, if to be transported by air, shall meet the requirements of para. 432 and shall not contain activities greater than the following:

(a) For *low dispersible radioactive material* — as authorized for the *package design* as specified in the certificate of *approval*;

(b) For *special form radioactive material* — $3000A_1$ or 10^5A_2, whichever is the lower;

(c) For all other *radioactive material* — $3000A_2$.

SPECIAL ARRANGEMENT

434. *Radioactive material* shall be classified as transported under *special arrangement* when it is intended to be carried in accordance with para. 310.

Section V

REQUIREMENTS AND CONTROLS FOR TRANSPORT

REQUIREMENTS BEFORE THE FIRST SHIPMENT

501. Before *packaging* is first used to transport *radioactive material*, it shall be confirmed that it has been manufactured in conformity with the *design* specifications to ensure compliance with the relevant provisions of these Regulations and any applicable certificate of *approval*. The following requirements shall also be fulfilled, if applicable:

(a) If the *design* pressure of the *containment system* exceeds 35 kPa (gauge), it shall be ensured that the *containment system* of each *packaging* conforms to the approved *design* requirements relating to the capability of that system to maintain its integrity under that pressure.
(b) For each *packaging* intended for use as a *Type B(U), Type B(M)* or *Type C package* and for each *packaging* intended to contain *fissile material*, it shall be ensured that the effectiveness of the shielding and containment and, where necessary, the heat transfer characteristics are within the limits applicable to or specified for the approved *design*.
(c) For each *packaging* intended to contain *fissile material*, it shall be ensured that the effectiveness of the criticality safety features is within the limits applicable to or specified for the *design*, and in particular where, in order to comply with the requirements of para. 673, neutron poisons are specifically included, checks shall be performed to confirm the presence and distribution of those neutron poisons.

REQUIREMENTS BEFORE EACH SHIPMENT

502. Before each *shipment* of a *package*, it shall be ensured that the *package* contains neither:

(a) Radionuclides different from those specified for the *package design*; nor
(b) Contents in a form, or physical or chemical state, different from those specified for the *package design*.

503. Before each *shipment* of a *package*, it shall be ensured that all the requirements specified in the relevant provisions of these Regulations and

in the applicable certificates of *approval* have been fulfilled. The following requirements shall also be fulfilled, if applicable:

(a) It shall be ensured that lifting attachments that do not meet the requirements of para. 608 have been removed or otherwise rendered incapable of being used for lifting the *package*, in accordance with para. 609.
(b) Each *Type B(U)*, *Type B(M)* and *Type C package* shall be held until equilibrium conditions have been approached closely enough to demonstrate compliance with the requirements for temperature and pressure, unless an exemption from these requirements has received *unilateral approval*.
(c) For each *Type B(U)*, *Type B(M)* and *Type C package*, it shall be ensured by inspection and/or appropriate tests that all closures, valves and other openings of the *containment system* through which the *radioactive contents* might escape are properly closed and, where appropriate, sealed in the manner for which the demonstrations of compliance with the requirements of paras 659 and 671 were made.
(d) For *packages* containing *fissile material*, the measurement specified in para. 677(b) and the tests to demonstrate closure of each *package* as specified in para. 680 shall be performed.
(e) For *packages* intended to be used for *shipment* after storage, it shall be ensured that all *packaging* components and *radioactive contents* have been maintained during storage in a manner such that all the requirements specified in the relevant provisions of these Regulations and in the applicable certificates of *approval* have been fulfilled.

TRANSPORT OF OTHER GOODS

504. A *package* shall not contain any items other than those that are necessary for the use of the *radioactive material*. The interaction between these items and the *package*, under the conditions of transport applicable to the *design*, shall not reduce the safety of the *package*.

505. *Freight containers*, *IBCs*, *tanks*, as well as other *packagings* and *overpacks*, used for the transport of *radioactive material* shall not be used for the storage or transport of other goods unless decontaminated below the level of 0.4 Bq/cm^2 for beta and gamma emitters and *low toxicity alpha emitters* and 0.04 Bq/cm^2 for all other alpha emitters.

506. *Consignments* shall be segregated from other dangerous goods during transport in compliance with the relevant transport regulations for dangerous

goods of each of the countries *through or into* which the materials will be transported, and, where applicable, with the regulations of the cognizant transport organizations, as well as these Regulations.

OTHER DANGEROUS PROPERTIES OF CONTENTS

507. In addition to the radioactive and fissile properties, any other dangerous properties of the contents of the *package*, such as explosiveness, flammability, pyrophoricity, chemical toxicity and corrosiveness, shall be taken into account in the classification, packing, labelling, marking, placarding, storage and transport in order to be in compliance with the relevant transport regulations for dangerous goods of each of the countries *through or into* which the materials will be transported, and, where applicable, with the regulations of the relevant transport organizations, as well as these Regulations.

REQUIREMENTS AND CONTROLS FOR CONTAMINATION AND FOR LEAKING PACKAGES

508. The *non-fixed contamination* on the external surfaces of a *package* shall be kept as low as practicable and, under routine conditions of transport, shall not exceed the following limits:

(a) 4 Bq/cm^2 for beta and gamma emitters and *low toxicity alpha emitters*;
(b) 0.4 Bq/cm^2 for all other alpha emitters.

These limits are applicable when averaged over any area of 300 cm^2 of any part of the surface.

509. Except as provided in para. 514, the level of *non-fixed contamination* on the external and internal surfaces of *overpacks*, *freight containers* and *conveyances* shall not exceed the limits specified in para. 508. This requirement does not apply to the internal surfaces of *freight containers* being used as *packaging*, either loaded or empty.

510. If it is evident that a *package* is leaking or damaged, or if it is suspected that the *package* may have leaked or been damaged, access to the *package* shall be restricted and a qualified person shall, as soon as possible, assess the extent of *contamination* and the resulting *dose rate* of the *package*. The scope of the assessment shall include the *package*, the *conveyance*, the adjacent loading and

unloading areas and, if necessary, all other material that has been carried in the *conveyance*. When necessary, additional steps for the protection of people, property and the environment, in accordance with provisions established by the relevant *competent authority*, shall be taken to overcome and minimize the consequences of such leakage or damage.

511. *Packages* that are damaged or leaking *radioactive contents* in excess of allowable limits for normal conditions of transport may be removed to an acceptable interim location under supervision, but shall not be forwarded until repaired or reconditioned and decontaminated.

512. A *conveyance* and equipment used regularly for the transport of *radioactive material* shall be periodically checked to determine the level of *contamination*. The frequency of such checks shall be related to the likelihood of *contamination* and the extent to which *radioactive material* is transported.

513. Except as provided in para. 514, any *conveyance*, or equipment or part thereof that has become contaminated above the limits specified in para. 508 in the course of the transport of *radioactive material*, or that shows a *dose rate* in excess of 5 μSv/h at the surface, shall be decontaminated as soon as possible by a qualified person and shall not be reused unless the following conditions are fulfilled:

(a) The *non-fixed contamination* shall not exceed the limits specified in para. 508.
(b) The *dose rate* resulting from the *fixed contamination* shall not exceed 5 μSv/h at the surface.

514. A *freight container* or *conveyance* dedicated to the transport of unpackaged *radioactive material* under *exclusive use* shall be excepted from the requirements of paras 509 and 513 solely with regard to its internal surfaces and only for as long as it remains under that specific *exclusive use*.

REQUIREMENTS AND CONTROLS FOR TRANSPORT OF EXCEPTED PACKAGES

515. *Excepted packages* shall be subject only to the following provisions in Sections V and VI:

(a) The requirements specified in paras 503–505, 507–513, 516, 530–533, 545, 546 (introductory sentence), 546(a), 546(j)(i) and (j)(ii), 546(k), 546(m), 550–553, 555, 556, 561, 564, 570, 582 and 583;
(b) The requirements for *excepted packages* specified in para. 622;
(c) The requirements specified in paras 580 and 581, if transported by post;
(d) The requirements specified in para. 536A, if the *package* is marked with a *package* type.

All relevant provisions of the other sections shall apply to *excepted packages*.

516. The *dose rate* at any point on the external surface of an *excepted package* shall not exceed 5 μSv/h.

REQUIREMENTS AND CONTROLS FOR TRANSPORT OF LSA MATERIAL AND SCO IN INDUSTRIAL PACKAGES OR UNPACKAGED

517. The quantity of *LSA material* and *SCO* in a single *Type IP-1, Type IP-2* or *Type IP-3 package*, or unpackaged *LSA material* and *SCO*, whichever is appropriate, shall be so restricted that the sum of the external *dose rates* at 3 m from the following unshielded items does not exceed 10 mSv/h:

(a) The *LSA material* (for a single group of *LSA material*) or each group of *LSA material* (when different groups of *LSA material* are packed together);
(b) The *SCO* (for a single *SCO*) or each *SCO* in a collection of *SCOs* (when different *SCOs* are packed together);
(c) Each group of *LSA material* and each *SCO* (when different groups of *LSA material* and *SCOs* are packed together).

518. For *LSA material* and *SCO* that are, or contain, *fissile material* that is not excepted under para. 417, the applicable requirements of paras 568 and 569 shall be met.

519. For *LSA material* and *SCO* that are, or contain, *fissile material*, the applicable requirements of para. 673 shall be met.

520. *LSA-I material* and *SCO-I* may be transported unpackaged, and *SCO-III* shall be transported unpackaged, under the following conditions:

(a) All unpackaged material other than ores containing only *naturally occurring radionuclides* shall be transported in such a manner that under routine conditions of transport there will be no escape of the *radioactive contents* from the *conveyance,* nor will there be any loss of shielding.
(b) Each *conveyance* or *large freight container* shall be under *exclusive use*, except when only transporting *SCO-I* on which the *contamination* on the accessible and the inaccessible surfaces is not greater than 10 times the applicable level specified in para. 214.
(c) For *SCO-I* where it is suspected that *non-fixed contamination* exists on inaccessible surfaces in excess of the values specified in para. 413(a)(i), measures shall be taken to ensure that the *radioactive material* is not released into the *conveyance*.
(d) Unpackaged *fissile material* shall meet the requirement of para. 417(e).
(e) For *SCO-III*:
    (i) The *conveyance* shall be under *exclusive use* by road, rail, inland waterway or sea.
    (ii) Stacking shall not be permitted.
    (iii) All activities associated with the *shipment*, including radiation protection, emergency response and any special precautions or special administrative or operational controls that are to be employed during transport shall be described in a transport plan. The transport plan shall demonstrate that the overall level of safety in transport is at least equivalent to that which would be provided if the requirements of para. 648 (only for the test specified in para. 724, preceded by the tests specified in paras 720 and 721) had been met.
    (iv) The requirements of para. 624 for a *Type IP-2 package* shall be satisfied, except that the maximum damage referred to in para. 722 may be determined based on provisions in the transport plan, and the requirements of para. 723 are not applicable.
    (v) The object and any shielding are secured to the *conveyance* in accordance with para. 607.
    (vi) The *shipment* shall be subject to *multilateral approval*.

521. *LSA material* and *SCO*, except as otherwise specified in para. 520, shall be packaged in accordance with Table 5. When different groups of *LSA material* and

TABLE 5. INDUSTRIAL PACKAGE REQUIREMENTS FOR LSA MATERIAL, SCO-I AND SCO-II

Radioactive contents	*Industrial package* type[a]	
	Exclusive use	Not under *exclusive use*
LSA-I		
Solid[b]	*Type IP-1*	*Type IP-1*
Liquid	*Type IP-1*	*Type IP-2*
LSA-II		
Solid	*Type IP-2*	*Type IP-2*
Liquid and gas	*Type IP-2*	*Type IP-3*
LSA-III	*Type IP-2*	*Type IP-3*
SCO-I[b]	*Type IP-1*	*Type IP-1*
SCO-II	*Type IP-2*	*Type IP-2*

[a] An *industrial package* of a higher type may be used (where *Type IP-1* is the lowest and *Type IP-3* is the highest; see paras 623–630).

[b] Under the conditions specified in para. 520, *LSA-I material* and *SCO-I* may be transported unpackaged.

SCO are packed in the same *industrial package*, and when those different groups satisfy the condition for different types of *industrial package*, the material shall be transported in the higher type of *industrial package* (where *Type IP-1* is the lowest and *Type IP-3* is the highest).

522. The total activity in a single hold or compartment of an inland waterway craft, or in another *conveyance*, for carriage of *LSA material* or *SCO* in a *Type IP-1, Type IP-2, Type IP-3 package* or unpackaged, shall not exceed the limits shown in Table 6. When different groups of *LSA material* and *SCO* are carried in the same hold or compartment of an inland waterway craft, or in another *conveyance*, the total activity in the hold or compartment, or in the other *conveyance*, shall not exceed the lowest appropriate limit indicated in Table 6.

TABLE 6. CONVEYANCE ACTIVITY LIMITS FOR LSA MATERIAL AND SCO IN INDUSTRIAL PACKAGES OR UNPACKAGED

Nature of material	Activity limit for *conveyances* other than inland waterway craft	Activity limit for a hold or compartment of an inland waterway craft
LSA-I	No limit	No limit
LSA-II and *LSA-III* non-combustible solids	No limit	$100A_2$
LSA-II and *LSA-III* combustible solids and all liquids and gases	$100A_2$	$10A_2$
SCO [a]	$100A_2$	$10A_2$

[a] For *SCO-III* see para. 522.

For *SCO-III*, the limits in Table 6 may be exceeded provided that the transport plan contains precautions that are to be employed during transport to obtain an overall level of safety at least equivalent to that which would be obtained if the limits had been applied.

DETERMINATION OF THE TRANSPORT INDEX

523. The *TI* for a *package,* a *tank* or a *freight container* used as *packaging*, or for unpackaged *LSA-I, SCO-I* or *SCO-III*, shall be the number derived in accordance with the following procedure:

(a) The maximum *dose rate* shall be determined in units of millisieverts per hour (mSv/h) at a distance of 1 m from the external surfaces of the *package*, the *tank* or the *freight container* used as *packaging*, or of the unpackaged *LSA-I, SCO-I* and *SCO-III*. The value determined shall be multiplied by 100. Alternatively, for uranium and thorium ores and their concentrates, the maximum *dose rate* at any point 1 m from the external surface of the load may be taken as:
    - (i) 0.4 mSv/h for ores and physical concentrates of uranium and thorium;
    - (ii) 0.3 mSv/h for chemical concentrates of thorium;
    - (iii) 0.02 mSv/h for chemical concentrates of uranium, other than uranium hexafluoride.

TABLE 7. MULTIPLICATION FACTORS FOR TANKS, FREIGHT CONTAINERS AND UNPACKAGED LSA-I, SCO-I AND SCO-III

Size of load[a]	Multiplication factor
size of load ≤ 1 m^2	1
1 m^2 < size of load ≤ 5 m^2	2
5 m^2 < size of load ≤ 20 m^2	3
20 m^2 < size of load	10

[a] Largest cross-sectional area of the load being measured.

(b) For a *tank* or a *freight container* used as *packaging*, and for unpackaged *LSA-I*, *SCO-I* and *SCO-III*, the value determined in step (a) shall be multiplied by the appropriate factor from Table 7.

(c) The value obtained in steps (a) and (b) shall be rounded up to the first decimal place (e.g. 1.13 shall be rounded up to 1.2), unless it is 0.05 or less, which may be considered as zero. The resulting number is the *TI* value.

524. The *TI* for each *overpack*, each *freight container* loaded with *packages* or each *conveyance* shall be determined as the sum of the *TIs* of all the *packages* contained therein. Alternatively, for a *shipment* from a single *consignor*, the *consignor* may determine the *TI* of a rigid *overpack,* a *freight container* loaded with *packages* or a *conveyance* in accordance with the following procedure:

(a) The maximum *dose rate* shall be determined in units of millisieverts per hour (mSv/h) at a distance of 1 m from the external surfaces of the rigid *overpack,* the *freight container* loaded with *packages* or the *conveyance.* The value determined shall be multiplied by 100. Alternatively, for uranium and thorium ores and their concentrates, the maximum *dose rate* at any point 1 m from the external surface of the load may be taken as:
   (i) 0.4 mSv/h for ores and physical concentrates of uranium and thorium;
   (ii) 0.3 mSv/h for chemical concentrates of thorium;
   (iii) 0.02 mSv/h for chemical concentrates of uranium other than uranium hexafluoride.

(b) For a *freight container* loaded with *packages*, the value determined in step (a) shall be multiplied by the appropriate factor from Table 7.
(c) The value obtained in steps (a) and (b) shall be rounded up to the first decimal place (e.g. 1.13 shall be rounded up to 1.2), unless it is 0.05 or less, which may be considered as zero. The resulting number is the *TI* value.

524A. This paragraph has been deleted and its content has been incorporated in para. 524.

DETERMINATION OF THE CRITICALITY SAFETY INDEX FOR CONSIGNMENTS, FOR FREIGHT CONTAINERS LOADED WITH PACKAGES AND FOR OVERPACKS

525. The *CSI* for each *overpack* or for each *freight container* loaded with *packages* shall be determined as the sum of the *CSIs* of all the *packages* contained. The same procedure shall be followed for determining the total sum of the *CSIs* in a *consignment* or aboard a *conveyance*.

LIMITS ON THE TRANSPORT INDEX, THE CRITICALITY SAFETY INDEX AND THE DOSE RATES FOR PACKAGES, FOR FREIGHT CONTAINERS USED AS PACKAGING AND FOR OVERPACKS

526. For *packages,* for *freight containers* used as *packaging* and for *overpacks,* other than those transported under *exclusive use*, the following shall apply:

(a) The *TI* of any *package,* other than a *freight container* used as *packaging,* or any *overpack* shall not exceed 10.
(b) The *TI* of any *freight container* used as *packaging* shall not exceed 50.
(c) The *CSI* of any *package*, any *freight container* used as *packaging* or any *overpack* shall not exceed 50.

527. The maximum *dose rate* at any point on the external surface of a *package*, a *freight container* used as *packaging* or an *overpack* shall not exceed 2 mSv/h, except for *packages*, *freight containers* used as *packaging* or *overpacks* transported under the following:

(a) *Exclusive use* by rail or by road, under the conditions specified in para. 573(a);
(b) *Exclusive use* or *special arrangement* by *vessel,* under the conditions specified in para. 575;
(c) *Special arrangement* by air, under the conditions specified in para. 579.

528. The maximum *dose rate* at any point on the external surface of a *package*, a *freight container* used as *packaging* or an *overpack* under *exclusive use* shall not exceed 10 mSv/h.

CATEGORIES

529. *Packages*, *overpacks* and *freight containers* shall be assigned to either category I-WHITE, II-YELLOW or III-YELLOW in accordance with the conditions specified in Table 8 and with the following requirements:

(a) For a *package, overpack* or *freight container*, the *TI* and the surface *dose rate* conditions shall be taken into account in determining which category is appropriate. For a *freight container* carrying unpackaged *LSA-I, SCO-I* or *SCO-III*, the *TI* and the surface *dose rate* to be taken into account are those of the unpackaged *LSA-I, SCO-I* or *SCO-III.* Where the *TI* satisfies the condition for one category but the surface *dose rate* satisfies the condition for a different category, the *package, overpack* or *freight container* shall be assigned to the higher category. For this purpose, category I-WHITE shall be regarded as the lowest category and III-YELLOW as the highest category.
(b) The *TI* shall be determined following the procedures specified in paras 523 and 524.
(c) If the surface *dose rate* is greater than 2 mSv/h, the *package* or *overpack* shall be transported under *exclusive use* and under the provisions of para. 573(a), 575 or 579, as appropriate.
(d) A *package* transported under a *special arrangement* shall be assigned to category III-YELLOW except under the provisions of para. 530.
(e) An *overpack* or *freight container* that contains *packages* transported under *special arrangement* shall be assigned to category III-YELLOW except under the provisions of para. 530.

TABLE 8. CATEGORIES OF PACKAGES, OVERPACKS AND FREIGHT CONTAINERS

Conditions		Category
TI	Maximum *dose rate* at any point on external surface	
0[a]	Not more than 0.005 mSv/h	I-WHITE
More than 0 but not more than 1[a]	More than 0.005 mSv/h but not more than 0.5 mSv/h	II-YELLOW
More than 1 but not more than 10	More than 0.5 mSv/h but not more than 2 mSv/h	III-YELLOW
More than 10	More than 2 mSv/h but not more than 10 mSv/h	III-YELLOW[b]

[a] If the measured *TI* is not greater than 0.05, the value quoted may be zero in accordance with para. 523(c).

[b] This category shall also be transported under *exclusive use* except for *freight containers* (see Table 10).

MARKING, LABELLING AND PLACARDING

530. For each *package* or *overpack*, the UN number and proper shipping name shall be determined (see Table 1). In all cases of international transport of *packages* requiring *competent authority approval* of the *design* or *shipment*, for which different *approval* types apply in the different countries concerned by the *shipment*, the UN number, proper shipping name, categorization, labelling and marking shall be in accordance with the certificate of the country of origin of the *design*.

Marking

531. Each *package* shall be legibly and durably marked on the outside of the *packaging* with an identification of either the *consignor* or *consignee*, or both. Each *overpack* shall be legibly and durably marked on the outside of the *overpack* with an identification of either the *consignor* or *consignee*, or both, unless the marks of all the *packages* within the *overpack* are clearly visible.

TABLE 9. UN MARKING FOR PACKAGES AND OVERPACKS

Item	UN marks[a]
Package (other than an *excepted package*)	UN number, preceded by the letters “UN”, and the proper shipping name for each applicable UN number in the *package*
Excepted package (other than a *package* accepted for international movement by post)	UN number, preceded by the letters “UN”
Overpack (other than an *overpack* containing only *excepted packages*)	UN number, preceded by the letters “UN” and followed by the proper shipping name in the case of a *non-excepted package*, for each applicable UN number in the *overpack*
Overpack containing only *excepted packages* (other than *packages* accepted for international movement by post)	UN number, preceded by the letters “UN”, for each applicable UN number in the *overpack*
Package accepted for international movement by post	The requirement of para. 581

[a] See Table 1 for a listing of UN numbers and proper shipping names.

532. Each *package* shall be legibly and durably marked on the outside with the UN marks as specified in Table 9. When two or more UN numbers apply to the *radioactive contents* within the same *packaging*, the *package* shall be marked as required for each applicable UN number. Additionally, each *overpack* shall be legibly and durably marked with the word “OVERPACK” and the UN marks as specified in Table 9, unless all the marks of the *packages* within the *overpack* are clearly visible.

533. Each *package* of gross mass exceeding 50 kg shall have its permissible gross mass legibly and durably marked on the outside of the *packaging*.

534. Each *package* that conforms to:

(a) An *IP-1*, *IP-2* or *IP-3 design* shall be legibly and durably marked on the outside of the *packaging* with “TYPE IP-1”, “TYPE IP-2” or “TYPE IP-3”, as appropriate.
(b) A *Type A package design* shall be legibly and durably marked on the outside of the *packaging* with “TYPE A”.

(c) An *IP-2*, *IP-3* or a *Type A package design* shall be legibly and durably marked on the outside of the *packaging* with the international *vehicle* registration code (VRI code) of the country of origin of the *design* and either the name of the manufacturer or other identification of the *packaging* specified by the *competent authority* of the country of origin of the *design*.

535. Each *package* that conforms to a *design* approved under one or more of paras 807–816 and 820 shall be legibly and durably marked on the outside of the *packaging* with the following information:

(a) The identification mark allocated to that *design* by the *competent authority*;
(b) A serial number to uniquely identify each *packaging* that conforms to that *design*;
(c) “TYPE B(U)”, “TYPE B(M)” or “TYPE C”, in the case of a *Type B(U)*, *Type B(M)* or *Type C package design*.

536. Each *package* that conforms to a *Type B(U)*, *Type B(M)* or *Type C package design* shall have the outside of the outermost receptacle that is resistant to the effects of fire and water plainly marked by embossing, stamping or other means resistant to the effects of fire and water with the trefoil symbol shown in Fig. 1.

536A. Any *package* type mark listed in para. 534(a) or (b) or in para. 535(c) that does not relate to the UN number and proper shipping name assigned shall be removed from the outside of the *package* or shall be covered.

537. Where *LSA-I material* or *SCO-I* is contained in receptacles or wrapping materials and is transported under *exclusive use*, as permitted by para. 520, the outer surface of these receptacles or wrapping materials may bear the mark “RADIOACTIVE LSA-I” or “RADIOACTIVE SCO-I”, as appropriate.

Labelling

538. Each *package*, *overpack* and *freight container* shall bear the labels conforming to the applicable models in Figs 2–4, except as allowed under the alternative provisions of para. 543 for *tanks* and *large freight containers*, in accordance with the appropriate category. In addition, each *package*, *overpack* and *freight container* containing *fissile material*, other than *fissile material* excepted under the provisions of para. 417, shall bear labels conforming to the model in Fig. 5. Any labels that do not relate to the contents shall be removed or covered. For *radioactive material* having other dangerous properties, see para. 507.

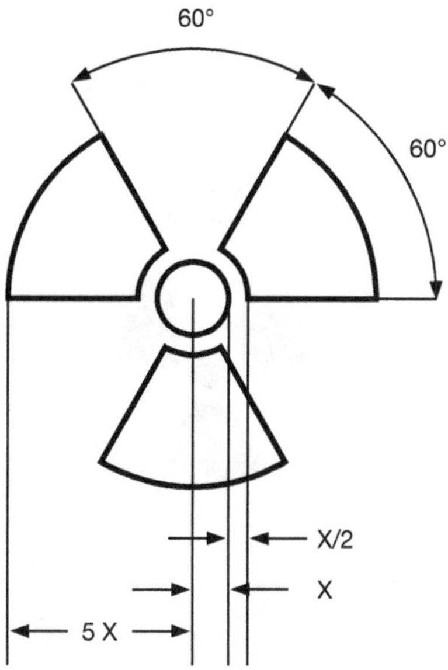

FIG. 1. Basic trefoil symbol with proportions based on a central circle of radius X. The minimum allowable size of X shall be 4 mm.

539. The labels conforming to the applicable models in Figs 2–4 shall be affixed to the two opposite sides of the outside of a *package* or *overpack* or to all four sides of the outside of a *tank* or *freight container* (even when used as *packaging*). The labels conforming to the model in Fig. 5, where applicable, shall be affixed adjacent to the labels conforming to the applicable models in Figs 2–4. The labels shall not cover the marks specified in paras 531–536.

Labelling for radioactive contents

540. Each label conforming to the applicable models in Figs 2–4 shall be completed with the following information:

(a) Contents:
   (i) Except for *LSA-I material*, the name(s) of the radionuclide(s) as taken from the Appendix, using the symbols prescribed therein. For mixtures of radionuclides, the most restrictive nuclides must be listed to the extent that the space on the line permits.
   (ii) For *LSA-I material*, the term *LSA-I* is all that is necessary; the name of the radionuclide is not necessary.

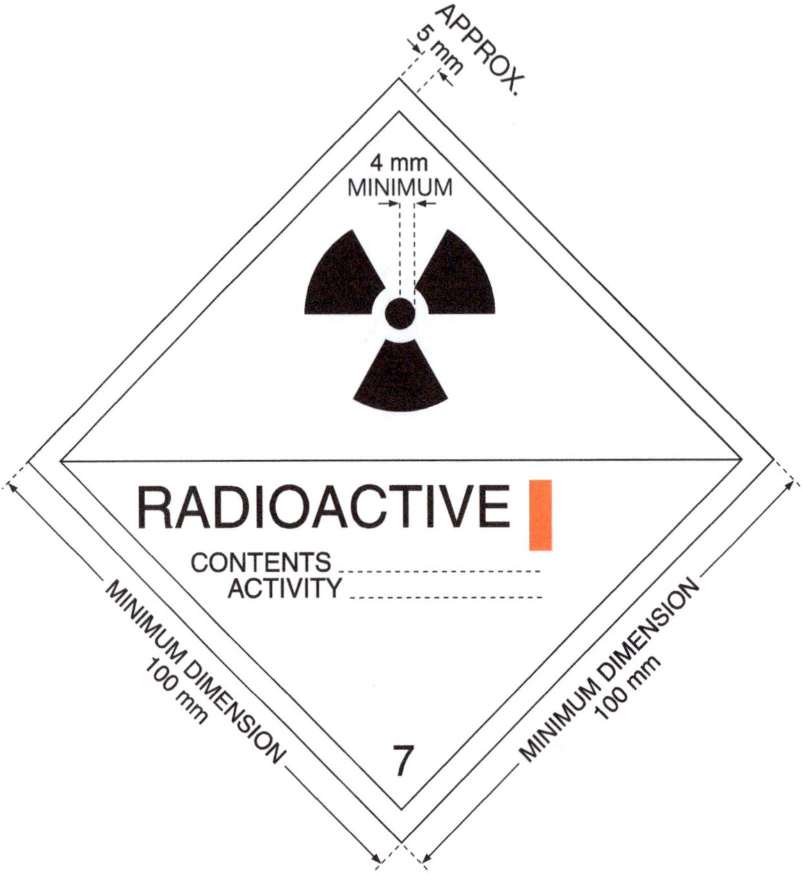

FIG. 2. Category I-WHITE label. The background colour of the label shall be white, the colour of the trefoil and the printing shall be black, and the colour of the category bar shall be red.

(b) Activity: The maximum activity of the *radioactive contents* during transport expressed in units of becquerels (Bq) with the appropriate prefix symbol of the International System of Units (SI) (see Annex II). For *fissile material*, the total mass of *fissile nuclides* in units of grams (g), or multiples thereof, may be used in place of activity.

(c) For *overpacks* and *freight containers*, the "CONTENTS" and "ACTIVITY" entries on the label shall bear the information required in paras 540(a) and (b), respectively, totalled together for the entire contents of the *overpack* or *freight container,* except that on labels for *overpacks* or *freight containers* containing mixed loads of *packages* containing different radionuclides, such entries may read "See Transport Documents".

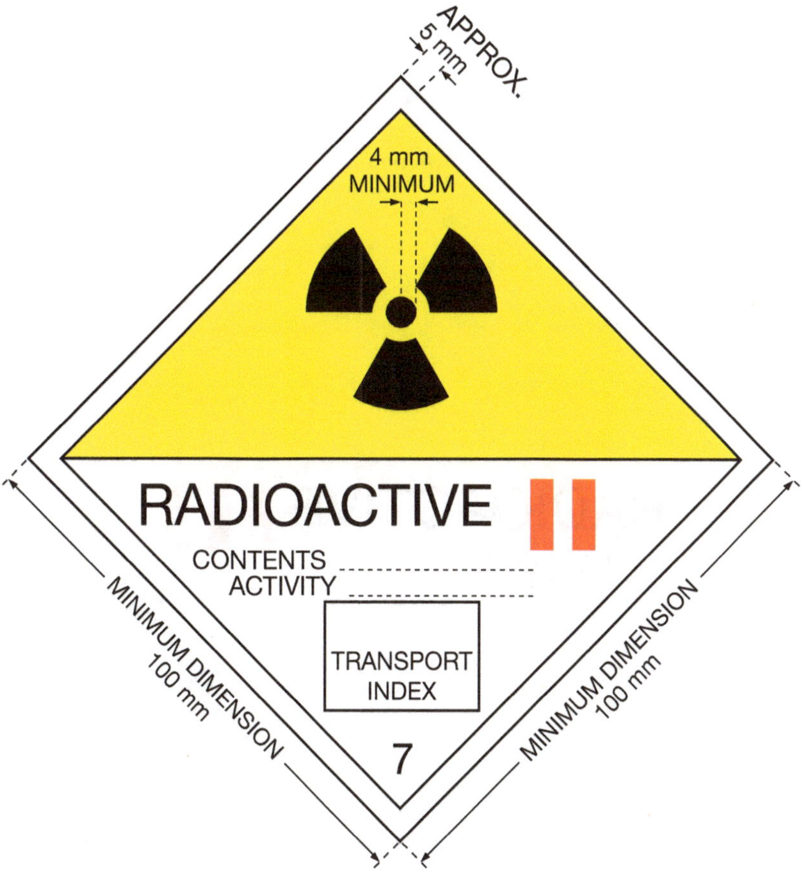

FIG. 3. Category II-YELLOW label. The background colour of the upper half of the label shall be yellow and the lower half white, the colour of the trefoil and the printing shall be black, and the colour of the category bars shall be red.

(d) *TI*: The number determined in accordance with paras 523 and 524 (except for category I-WHITE).

Labelling for criticality safety

541. Each label conforming to the model in Fig. 5 shall be completed with the *CSI* as stated in the certificate of *approval* applicable in the countries *through or into* which the *package* is transported and issued by the *competent authority* or as specified in para. 674 or 675.

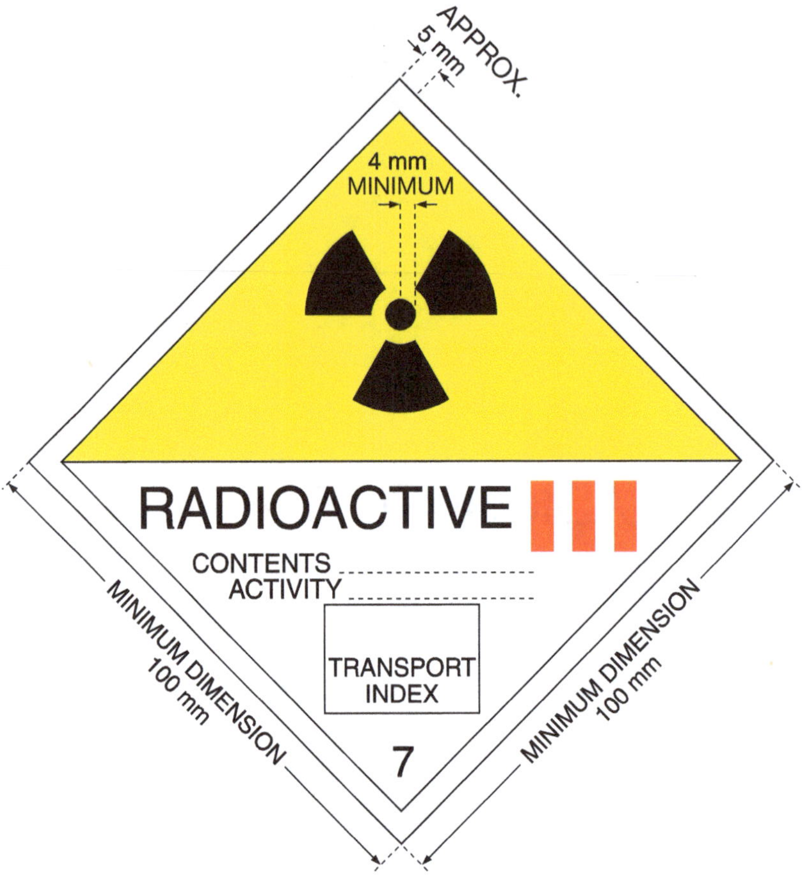

FIG. 4. Category III-YELLOW label. The background colour of the upper half of the label shall be yellow and the lower half white, the colour of the trefoil and the printing shall be black, and the colour of the category bars shall be red.

542. For *overpacks* and *freight containers* loaded with *packages*, the label conforming to the model in Fig. 5 shall bear the sum of the *CSIs* of all the *packages* contained therein.

Placarding

543. *Tanks* and *large freight containers* (other than *large freight containers* carrying only *excepted packages*) shall bear four placards that conform to the model given in Fig. 6. The placards shall be affixed in a vertical orientation to each side wall and to each end wall of the *tank* or *large freight container*. Any placards that do not relate to the contents shall be removed. Instead of using both

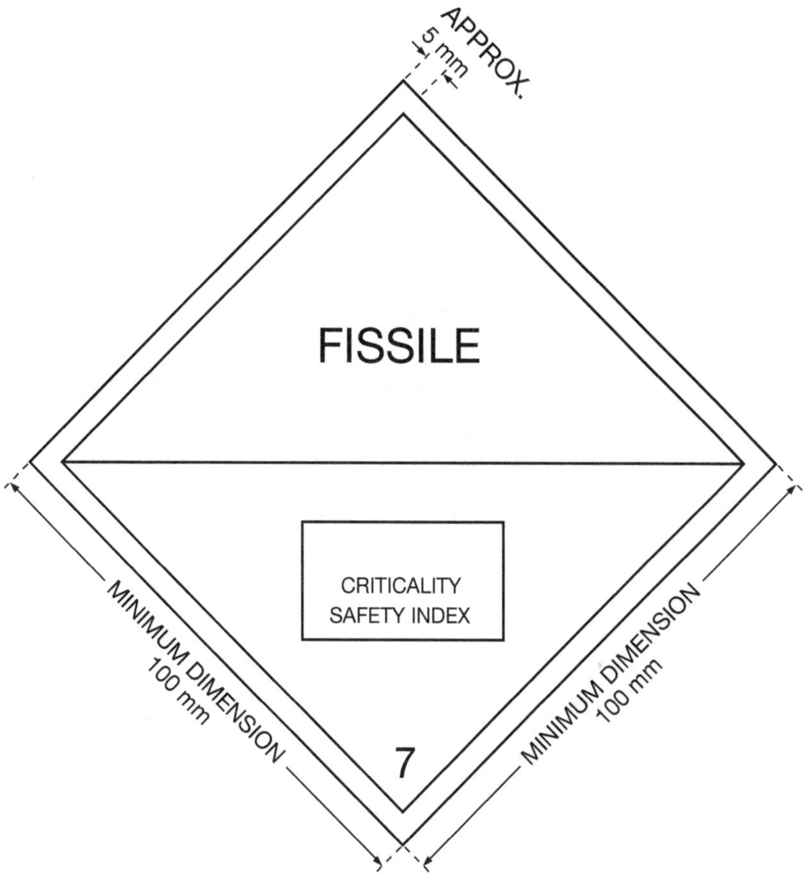

FIG. 5. CSI label. The background colour of the label shall be white, the colour of the printing shall be black.

labels and placards, it is permitted, as an alternative, to use enlarged labels only, where appropriate, as shown in Figs 2–4, but with the minimum dimensions shown in Fig. 6.

544. Where the *radioactive material* is in a *tank*, or is unpackaged *LSA-I* or *SCO-I* being carried in a *freight container*, or where a *consignment* in a *freight container* is required to be shipped under *exclusive use* and is packaged *radioactive material* with a single UN number, or is *SCO-III*, the appropriate

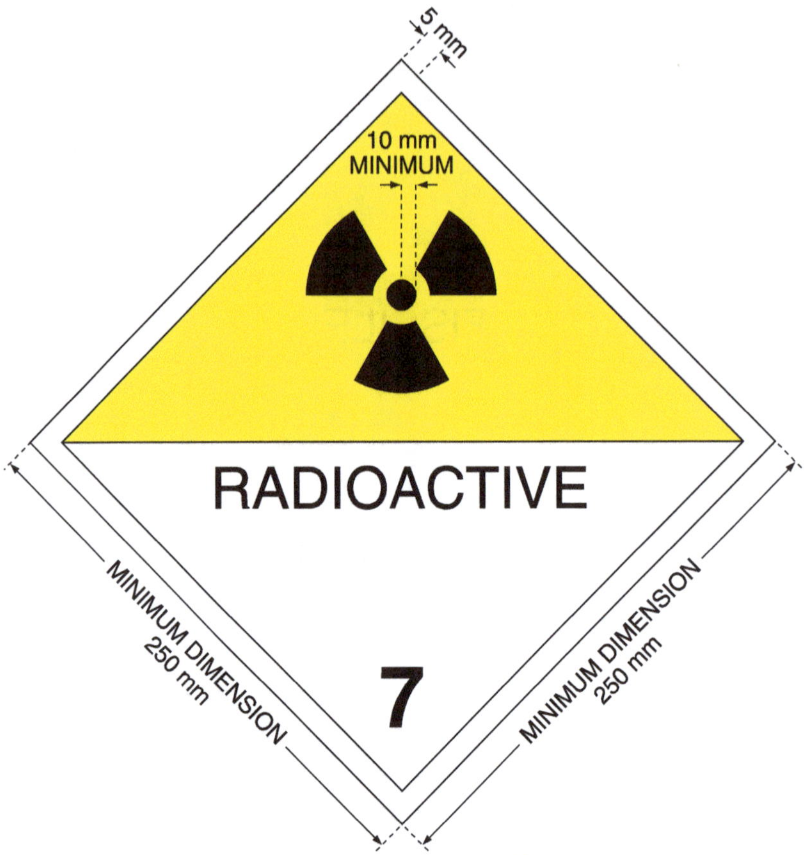

FIG. 6. Placard. Except as permitted by para. 571, minimum dimensions shall be as shown; when different dimensions are used, the relative proportions shall be maintained. The number '7' shall not be less than 25 mm high. The background colour of the upper half of the placard shall be yellow and of the lower half white, the colour of the trefoil and the printing shall be black. The use of the word "RADIOACTIVE" in the bottom half is optional, to allow the alternative use of this placard to display the appropriate UN number for the radioactive material.

UN number for the *radioactive material* (see Table 1) shall also be displayed, in black digits not less than 65 mm high, either:

(a) In the lower half of the placard shown in Fig. 6 and against the white background; or
(b) On the placard shown in Fig. 7.

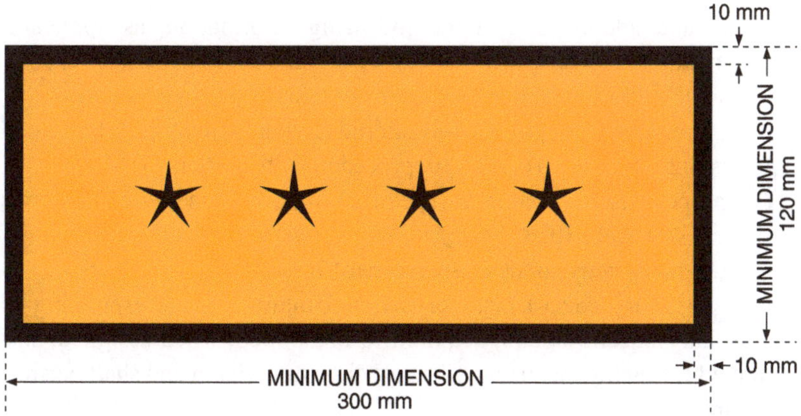

FIG. 7. Placard for separate display of UN number. The background colour of the placard shall be orange and the border and UN number shall be black. The symbol "★★★★" denotes the space in which the appropriate UN number for radioactive material, as specified in Table 1, shall be displayed.

When the alternative given in subparagraph (b) is used, the subsidiary placard shall be affixed immediately adjacent to the main placard shown in Fig. 6, on all four sides of the *tank, freight container* or *SCO-III*.

544A. Any empty *large freight container* for which the provisions of para. 514 are applied shall bear placards as required by paras 543 and 544 for unpackaged *LSA-I* or *SCO-I* previously transported in this *large freight container*.

CONSIGNOR'S RESPONSIBILITIES

545. Except as otherwise provided in these Regulations, no person may offer *radioactive material* for transport unless it is properly marked, labelled, placarded, described and certified on a transport document, and otherwise in a condition for transport as required by these Regulations.

Particulars of consignment

546. The *consignor* shall include in the transport document with each *consignment* the identification of the *consignor* and *consignee*, including

their names and addresses, and the following information, as applicable, in the order given:

(a) The UN number assigned to the material as specified in accordance with the provisions of paras 401 and 530, preceded by the letters “UN”.
(b) The proper shipping name, as specified in accordance with the provisions of paras 401 and 530.
(c) The UN dangerous goods class number “7”.
(d) The subsidiary hazard class or division number(s) corresponding to the subsidiary hazard label(s) required to be applied, when assigned, shall be entered following the primary hazard class or division and shall be enclosed in parentheses.
(e) The name or symbol of each radionuclide or, for mixtures of radionuclides, an appropriate general description or a list of the most restrictive nuclides.
(f) A description of the physical and chemical form of the material, or a notation that the material is *special form radioactive material* or *low dispersible radioactive material*. A generic chemical description is acceptable for the chemical form.
(g) The maximum activity of the *radioactive contents* during transport expressed in units of becquerels (Bq) with the appropriate SI prefix symbol (see Annex II). For *fissile material*, the mass of *fissile material* (or mass of each *fissile nuclide* for mixtures, when appropriate) in units of grams (g), or appropriate multiples thereof, may be used in place of activity.
(h) The category of the *package, overpack* or *freight container*, as assigned in compliance with para. 529 (i.e. I-WHITE, II-YELLOW, III-YELLOW).
(i) The *TI* as determined in accordance with paras 523 and 524 (except for category I-WHITE).
(j) For *fissile material*:
    (i) Meeting one of the provisions of subparagraphs (a)–(h) of para. 417, reference to that paragraph;
    (ii) Meeting one of the provisions of subparagraphs (c)–(e) of para. 417, the total mass of *fissile nuclides*;
    (iii) Contained in a *package* for which one of paras 674(a)–(c) or 675 is applied, reference to that paragraph;
    (iv) The *CSI*, where applicable.
(k) The identification mark for each *competent authority* certificate of *approval* (*special form radioactive material, low dispersible radioactive material, fissile material* excepted under para. 417(f), *special arrangement, package design* or *shipment*) applicable to the *consignment*.
(l) For *consignments* of more than one *package*, the information contained in paras 546(a)–(k) shall be given for each *package*. For *packages* in

an *overpack, freight container* or *conveyance*, a detailed statement of the contents of each *package* within the *overpack, freight container* or *conveyance* and, where appropriate, of each *overpack, freight container* or *conveyance* shall be included. If *packages* are to be removed from the *overpack, freight container* or *conveyance* at a point of intermediate unloading, appropriate transport documents shall be made available.

(m) Where *exclusive use* is required (see Annex III), the statement "EXCLUSIVE USE".

(n) For *LSA-II, LSA-III, SCO-I, SCO-II* and *SCO-III*, the total activity of the *consignment* as a multiple of A_2. For *radioactive material* for which the A_2 value is unlimited, the multiple of A_2 shall be zero.

Note: When two or more UN numbers apply to the *radioactive contents* within the same *packaging*, the information required by paras 546(a)–(f) shall be provided separately for each applicable UN number.

546A. When the provisions of para. 514 are applied to an empty *large freight container* or *conveyance*, the *consignment* shall be described as such by, for example, placing the words "EMPTY UNCLEANED" or "RESIDUE LAST CONTAINED" as appropriate, before or after the information required by paras 546(a)–(c) that were applicable to the last unpackaged *LSA-I* or *SCO-I* carried in the *large freight container* or *conveyance*, followed by the words "EXCLUSIVE USE".

Consignor's certification or declaration

547. The *consignor* shall include in the transport document a certification or declaration in the following terms:

> "I hereby declare that the contents of this consignment are fully and accurately described above by the proper shipping name, and are classified, packaged, marked and labelled/placarded, and are in all respects in proper condition for transport according to applicable international and national governmental regulations."

548. If the intent of the declaration is already a condition of transport within a particular international convention, the *consignor* need not provide a declaration for that part of the transport covered by the convention.

549. The declaration shall be signed and dated by the *consignor*. Facsimile signatures are acceptable where applicable laws and regulations recognize the legal validity of facsimile signatures.

550. If the dangerous goods documentation is presented to the *carrier* by means of electronic data processing (EDP) or electronic data interchange (EDI) transmission techniques, the signature(s) may be electronic signature(s) or may be replaced by the name(s) (in capitals) of the person authorized to sign.

551. When *radioactive material*, other than when carried in *tanks*, is packed or loaded into any *freight container* or *vehicle* that will be transported by sea, those responsible for packing the container or *vehicle* shall provide a container/*vehicle* packing certificate specifying the container/*vehicle* identification number(s) and certifying that the operation has been carried out in accordance with the applicable conditions of the International Maritime Dangerous Goods (IMDG) Code [18].

552. The information required in the transport document and the container/*vehicle* packing certificate may be incorporated into a single document, if not, the documents shall be attached. If the information is incorporated into a single document, the document shall include a signed declaration such as:

> "It is declared that the packing of the goods into the container/vehicle has been carried out in accordance with the applicable provisions".

This declaration shall be dated and the person signing it shall be identified on the document. Facsimile signatures are acceptable where applicable laws and regulations recognize the legal validity of facsimile signatures.

553. The declaration shall be made on the same transport document that contains the particulars of *consignment* listed in para. 546.

Information for carriers

554. The *consignor* shall provide in the transport document a statement regarding actions, if any, that are required to be taken by the *carrier*. The statement shall be in the languages deemed necessary by the *carrier* or the authorities concerned and shall include at least the following points:

(a) Supplementary requirements for loading, stowage, carriage, handling and unloading of the *package*, *overpack* or *freight container*, including specific directions when *exclusive use* is required and any special stowage provisions for the safe dissipation of heat (see para. 565), or a statement that no such requirements are necessary;

(b) Restrictions on the mode of transport or *conveyance* and any necessary routeing instructions;

(c) Emergency arrangements appropriate to the *consignment.*

555. The *consignor* shall retain a copy of each of the transport documents containing the information specified in paras 546, 547, 551, 552 and 554, as applicable, for a minimum period of three months.

When the documents are kept electronically, the *consignor* shall be able to reproduce them in a printed form.

556. The applicable *competent authority* certificates need not necessarily accompany the *consignment*. The *consignor* shall make the applicable certificates available to the *carrier(s)* before loading and unloading.

Notification of competent authorities

557. Before the first *shipment* of a *package* requiring *competent authority approval*, the *consignor* shall ensure that copies of each applicable *competent authority* certificate applying to that *package design* have been provided to the *competent authority* of the country of origin of the *shipment* and to the *competent authority* of each country *through or into* which the *consignment* is to be transported. The *consignor* is not required to wait for an acknowledgement from the *competent authority*, nor is the *competent authority* required to make such an acknowledgement of receipt of the certificate.

558. For each *shipment* listed in subparagraph (a), (b), (c) or (d) of this paragraph, the *consignor* shall notify the *competent authority* of the country of origin of the *shipment* and the *competent authority* of each country *through or into* which the *consignment* is to be transported. This notification shall be in the possession of each *competent authority* prior to the commencement of the *shipment*, preferably at least 7 days in advance of the *shipment*.

(a) *Type C packages* containing *radioactive material* with an activity greater than $3000A_1$ or $3000A_2$, as appropriate, or 1000 TBq, whichever is the lower;
(b) *Type B(U) packages* containing *radioactive material* with an activity greater than $3000A_1$ or $3000A_2$, as appropriate, or 1000 TBq, whichever is the lower;
(c) *Type B(M) packages*;
(d) *Shipments* under *special arrangement*.

559. The *consignment* notification shall include:

(a) Sufficient information to enable the identification of the *package* or *packages*, including all applicable certificate numbers and identification marks.
(b) Information on the date of *shipment*, the expected date of arrival and the proposed routeing.
(c) The name(s) of the *radioactive material(s)* or nuclide(s).
(d) Descriptions of the physical and chemical forms of the *radioactive material*, or whether it is *special form radioactive material* or *low dispersible radioactive material*.
(e) The maximum activity of the *radioactive contents* during transport expressed in units of becquerels (Bq) with the appropriate SI prefix symbol (see Annex II). For *fissile material*, the mass of *fissile material* (or the mass of each *fissile nuclide* for a mixture, when appropriate) in units of grams (g), or multiples thereof, may be used in place of activity.

560. The *consignor* is not required to send a separate notification if the required information has been included in the application for *approval* of *shipment* (see para. 827).

Possession of certificates and instructions

561. The *consignor* shall have in his/her possession a copy of each certificate required under Section VIII of these Regulations and a copy of the instructions with regard to the proper closing of the *package* and other preparations for *shipment* before making any *shipment* under the terms of the certificates.

TRANSPORT AND STORAGE IN TRANSIT

Segregation during transport and storage in transit

562. *Packages*, *overpacks* and *freight containers* containing *radioactive material* and unpackaged *radioactive material* shall be segregated during transport and during storage in transit:

(a) From workers in regularly occupied working areas, by distances calculated using a dose criterion of 5 mSv in a year and conservative model parameters;

(b) From members of the public in areas where the public has regular access, by distances calculated using a dose criterion of 1 mSv in a year and conservative model parameters;
(c) From undeveloped photographic film, by distances calculated using a radiation exposure criterion for undeveloped photographic film due to the transport of *radioactive material* of 0.1 mSv per *consignment* of such film;
(d) From other dangerous goods in accordance with para. 506.

563. Category II-YELLOW or III-YELLOW *packages* or *overpacks* shall not be carried in compartments occupied by passengers, except those exclusively reserved for couriers specially authorized to accompany such *packages* or *overpacks*.

Stowage during transport and storage in transit

564. *Consignments* shall be securely stowed.

565. Provided that its average surface heat flux does not exceed 15 W/m^2 and that the immediate surrounding cargo is not in sacks or bags, a *package* or *overpack* may be carried or stored among packaged general cargo without any special stowage provisions except as may be specifically required by the *competent authority* in an applicable certificate of *approval*.

566. Loading of *packages* and *overpacks* in *freight containers* and accumulation of *packages*, *overpacks* and *freight containers* aboard *conveyances* shall be controlled as follows:

(a) Except under the condition of *exclusive use*, and for *consignments* of *LSA-I material*, the total number of *packages* and *overpacks* in a *freight container* and of *packages*, *overpacks* and *freight containers* aboard a single *conveyance* shall be limited so that the sum of the *TIs* in the *freight container* and aboard the *conveyance*, respectively, does not exceed the values shown in Table 10.
(b) The *dose rate* under routine conditions of transport shall not exceed 2 mSv/h at any point on the external surface of the *vehicle* or *freight container*, and 0.1 mSv/h at 2 m from the external surface of the *vehicle* or *freight container*, except for *vehicles* or *large freight containers* under *exclusive use* by road or rail for which the *dose rate* limits around the *vehicle* are specified in paras 573(b) and (c).
(c) The sum of the *CSIs* in a *freight container* and aboard a *conveyance* shall not exceed the values shown in Table 11.

TABLE 10. TRANSPORT INDEX LIMITS FOR FREIGHT CONTAINERS AND CONVEYANCES NOT UNDER EXCLUSIVE USE

Type of *freight container* or *conveyance*	Limit on sum of *TIs* in a *freight container* or aboard a *conveyance*
Freight container loaded with *packages*[a] and *overpacks*:	
Small freight container	50
Large freight container	50
Vehicle[b]	50
Aircraft[b]:	
Passenger	50
Cargo	200
Inland waterway craft[b]	50
Sea-going *vessel*[c]:	
(i) Hold, compartment or *defined deck area*:	
Packages[d,e], *overpacks*[f], *small freight containers*	50
Large freight containers	200
(ii) Total *vessel*:	
Packages[d,e], *overpacks*[f], *small freight containers*	200
Large freight containers	No limit

[a] For *packages* contained in an *overpack*, the *TI* of the *overpack* shall be used to evaluate the sum of the *TIs*.

[b] For *packages* contained in an *overpack*, the *TI* of the *overpack* shall be used to evaluate the sum of the *TIs*. For *packages* or *overpacks* loaded in a *freight container*, the *TI* of the *freight container* shall be used to evaluate the sum of the *TIs*.

[c] *Packages* or *overpacks* carried in or on a *vehicle* that are in accordance with the provisions of para. 573 may be transported by *vessels* provided that they are not removed from the *vehicle* at any time while on board the *vessel*.

[d] For *freight containers* used as *packaging* and *packages* loaded in a *freight container*, the *TI* limit for either *small freight containers* or *large freight containers*, as appropriate, shall apply.

[e] For *packages* contained in an *overpack*, the *TI* limit for *overpacks* shall apply.

[f] For *overpacks* contained in a *freight container*, the *TI* limit for *small freight containers* or *large freight containers*, as appropriate, shall apply.

TABLE 11. CRITICALITY SAFETY INDEX LIMITS FOR FREIGHT CONTAINERS AND CONVEYANCES CONTAINING FISSILE MATERIAL

Type of *freight container* or *conveyance*	Limit on sum of *CSIs* in a *freight container* or aboard a *conveyance*	
	Not under *exclusive use*	Under *exclusive use* [a]
Freight container loaded with *packages*:		
Small freight container	50	Not applicable
Large freight container	50	100
Vehicle	50	100
Aircraft:		
Passenger	50	Not applicable
Cargo	50	100
Inland waterway craft	50	100
Sea-going *vessel*:		
(i) Hold, compartment or *defined deck area*	50	100
(ii) Total *vessel*:		
Packages, overpacks, small freight containers	200[b]	200[c]
Large freight containers	No limit[b]	No limit[c]

[a] Requires *shipment approval* in accordance with para. 825(c), as applicable.

[b] The *consignments* shall be so handled and stowed that the sum of *CSIs* in any group does not exceed 50 and that each group is handled and stowed so as to maintain a spacing of at least 6 m from other groups.

[c] The *consignments* shall be so handled and stowed that the sum of *CSIs* in any group does not exceed 100 and that each group is handled and stowed so as to maintain a spacing of at least 6 m from other groups or other *conveyances* carrying *radioactive material*. The intervening space between groups may be occupied by other cargo in accordance with para. 506.

567. This paragraph has been deleted.

Additional requirements relating to transport and storage in transit of fissile material

568. Any group of *packages*, *overpacks* and *freight containers* containing *fissile material* stored in transit in any one storage area shall be so limited that the sum of the *CSIs* in the group does not exceed 50. Each group shall be stored so as to maintain a spacing of at least 6 m from other such groups.

569. Where the sum of the *CSIs* on board a *conveyance* or in a *freight container* exceeds 50, as permitted in Table 11, storage shall be such as to maintain a spacing of at least 6 m from other groups of *packages*, *overpacks* or *freight containers* containing *fissile material* or other *conveyances* carrying *radioactive material.*

570. *Fissile material* meeting one of the provisions of subparagraphs (a)–(h) of para. 417 shall meet the following requirements:

(a) Only one of the provisions of subparagraphs (a)–(h) of para. 417 is allowed per *consignment.*
(b) Only one approved *fissile material* in *packages* classified in accordance with para. 417(f) is allowed per *consignment* unless multiple materials are authorized in the certificate of *approval.*
(c) *Fissile material* in *packages* classified in accordance with para. 417(c) shall be transported in a *consignment* with no more than 45 g of *fissile nuclides.*
(d) *Fissile material* in *packages* classified in accordance with para. 417(d) shall be transported in a *consignment* with no more than 15 g of *fissile nuclides.*
(e) Unpackaged or packaged *fissile material* classified in accordance with para. 417(e) shall be transported on a *conveyance* or in a *large freight container* under *exclusive use* with no more than 45 g of *fissile nuclides* on the *conveyance.*
(f) *Fissile material* in *packages* classified in accordance with para. 417(h) shall be transported in a *consignment* with no more than 45 g of *fissile nuclides.*

Additional requirements relating to transport by rail and by road

571. *Vehicles* carrying *packages*, *overpacks* or *freight containers* labelled with any of the labels shown in Figs 2–5, or carrying unpackaged *LSA-I material*, *SCO-I* or *SCO-III*, shall display the placard shown in Fig. 6 on each of:

(a) The two external lateral walls in the case of a rail *vehicle*;

(b) The two external lateral walls and the external rear wall in the case of a road *vehicle*.

In the case of a *vehicle* without sides, the placards may be affixed directly on the cargo carrying unit provided that they are readily visible. In the case of large *tanks* or *freight containers*, the placards on the *tanks* or *freight containers* shall suffice. In the case of *vehicles* that have insufficient area to allow the fixing of larger placards, the dimensions of the placard described in Fig. 6 may be reduced to 100 mm. Any placards that do not relate to the contents shall be removed.

572. Where the *radioactive material* in or on the *vehicle* is unpackaged *LSA-I material, SCO-I* or *SCO-III,* or where a *consignment* is required to be shipped under *exclusive use* and is packaged *radioactive material* with a single UN number, the appropriate UN number (see Table 1) shall also be displayed, in black digits not less than 65 mm high, either:

(a) In the lower half of the placard shown in Fig. 6, against the white background; or
(b) On the placard shown in Fig. 7.

When the alternative given in subparagraph (b) is used, the subsidiary placard shall be affixed immediately adjacent to the main placard, either on the two external lateral walls in the case of a rail *vehicle* or on the two external lateral walls and the external rear wall in the case of a road *vehicle.*

572A. Any empty *vehicle* for which the provisions of para. 514 are applied shall bear placards as required by paras 571 and 572 for unpackaged *LSA-I* or *SCO-I* previously transported in this *vehicle.*

573. For *vehicles* or *large freight containers* under *exclusive use*, the *dose rate* shall not exceed:

(a) 10 mSv/h at any point on the external surface of a *package,* a *freight container* used as *packaging* or an *overpack*, and may only exceed 2 mSv/h provided that:
   (i) The *vehicle* is equipped with an enclosure that, during routine conditions of transport, prevents the access of unauthorized persons to the interior of the enclosure.
   (ii) Provisions are made to secure the *package,* the *freight container* used as *packaging* or the *overpack* so that its position within the *vehicle* enclosure remains fixed during routine conditions of transport.

(iii) There is no loading or unloading during the *shipment*.

(b) 2 mSv/h at any point on the outer surfaces of the *vehicle*, including the upper and lower surfaces, or, in the case of an open *vehicle*, at any point on the vertical planes projected from the outer edges of the *vehicle*, on the upper surface of the load and on the lower external surface of the *vehicle*.

(c) 0.1 mSv/h at any point 2 m from the vertical planes represented by the outer lateral surfaces of the *vehicle*, or, if the load is transported in an open *vehicle*, at any point 2 m from the vertical planes projected from the outer edges of the *vehicle*.

574. In the case of road *vehicles*, no persons other than the driver and assistants shall be permitted in *vehicles* carrying *packages*, *overpacks* or *freight containers* bearing category II-YELLOW or III-YELLOW labels.

Additional requirements relating to transport by vessels

575. *Packages, freight containers* used as *packagings* or *overpacks* having a surface *dose rate* greater than 2 mSv/h, unless being carried in or on a *vehicle* under *exclusive use* in accordance with Table 10, footnote (c), shall not be transported by *vessel* except under *special arrangement*.

576. The transport of *consignments* by means of a special use *vessel* that, by virtue of its *design*, or by reason of its being chartered, is dedicated to the purpose of carrying *radioactive material*, shall be excepted from the requirements specified in para. 566 provided that the following conditions are met:

(a) A *radiation protection programme* for the *shipment* shall be approved by the *competent authority* of the flag State of the *vessel* and, when requested, by the *competent authority* at each port of call.

(b) Stowage arrangements shall be predetermined for the whole voyage, including any *consignments* to be loaded at ports of call en route.

(c) The loading, carriage and unloading of the *consignments* shall be supervised by persons qualified in the transport of *radioactive material*.

Additional requirements relating to transport by air

577. *Type B(M) packages* and *consignments* under *exclusive use* shall not be transported on passenger *aircraft*.

578. Vented *Type B(M) packages*, *packages* that require external cooling by an ancillary cooling system, *packages* subject to operational controls during

transport and *packages* containing liquid pyrophoric materials shall not be transported by air.

579. *Packages, freight containers* used as *packagings* or *overpacks* having a surface *dose rate* greater than 2 mSv/h shall not be transported by air except under *special arrangement.*

Additional requirements relating to transport by post

580. A *package* that conforms to the requirements of para. 515, in which the activity of the *radioactive contents* does not exceed one tenth of the limits prescribed in Table 4, and that does not contain uranium hexafluoride, may be accepted for domestic movement by national postal authorities, subject to such additional requirements as those authorities may prescribe.

581. A *package* that conforms to the requirements of para. 515, in which the activity of the *radioactive contents* does not exceed one tenth of the limits prescribed in Table 4, and that does not contain uranium hexafluoride, may be accepted for international movement by post, subject in particular to the following additional requirements as prescribed by the Acts of the Universal Postal Union:

(a) The *package* shall be deposited with the postal service only by *consignors* authorized by the national authority.
(b) The *package* shall be dispatched by the quickest route, normally by air.
(c) The *package* shall be plainly and durably marked on the outside with the words “RADIOACTIVE MATERIAL — QUANTITIES PERMITTED FOR MOVEMENT BY POST”. These words shall be crossed out if the *packaging* is returned empty.
(d) The *package* shall carry on the outside the name and address of the *consignor* with the request that the *package* be returned in the case of non-delivery.
(e) The name and address of the *consignor* and the contents of the *package* shall be indicated on the internal *packaging*.

CUSTOMS OPERATIONS

582. Customs operations involving the inspection of the *radioactive contents* of a *package* shall be carried out only in a place where adequate means of controlling radiation exposure are provided and in the presence of qualified persons. Any *package* opened on customs instructions shall, before being forwarded to the *consignee*, be restored to its original condition.

UNDELIVERABLE CONSIGNMENTS

583. Where a *consignment* is undeliverable, it shall be placed in a safe location and the appropriate *competent authority* shall be informed as soon as possible and a request made for instructions on further action.

RETENTION AND AVAILABILITY OF TRANSPORT DOCUMENTS BY CARRIERS

584. A *carrier* shall not accept a *consignment* for transport unless:

(a) A copy of the transport document and other documents or information as required by these Regulations are provided; or
(b) The information applicable to the *consignment* is provided in electronic form.

585. The information applicable to the *consignment* shall accompany the *consignment* to its final destination. This information may be on the transport document or may be on another document. This information shall be given to the *consignee* when the *consignment* is delivered.

586. When the information applicable to the *consignment* is given to the *carrier* in electronic form, the information shall be available to the *carrier* at all times during transport to the *consignment's* final destination. The information shall be able to be produced without delay in a printed form.

587. The *carrier* shall retain a copy of the transport document and additional information and documentation, as specified in these Regulations, for a minimum period of three months.

588. When the documents are kept electronically or in a computer system, the *carrier* shall be capable of reproducing them in a printed form.

Section VI

REQUIREMENTS FOR RADIOACTIVE MATERIAL AND FOR PACKAGINGS AND PACKAGES

REQUIREMENTS FOR RADIOACTIVE MATERIAL

Requirements for LSA-III material

601. This paragraph was deleted.

Requirements for special form radioactive material

602. *Special form radioactive material* shall have at least one dimension of not less than 5 mm.

603. *Special form radioactive material* shall be of such a nature or shall be so designed that if it is subjected to the tests specified in paras 704–711, it shall meet the following requirements:

(a) It would not break or shatter under the impact, percussion and bending tests in paras 705–707 and 709(a), as applicable.
(b) It would not melt or disperse in the heat test in para. 708 or 709(b), as applicable.
(c) The activity in the water from the leaching tests specified in paras 710 and 711 would not exceed 2 kBq; or alternatively, for sealed sources, the leakage rate for the volumetric leakage assessment test specified in the International Organization for Standardization document: Radiation Protection — Sealed Sources — Leakage Test Methods (ISO 9978) [19], would not exceed the applicable acceptance threshold acceptable to the *competent authority*.

604. When a sealed capsule constitutes part of the *special form radioactive material*, the capsule shall be so manufactured that it can be opened only by destroying it.

604A. The *design* of *special form radioactive material* shall take into account ageing mechanisms.

Requirements for low dispersible radioactive material

605. *Low dispersible radioactive material* shall be such that the total amount of this *radioactive material* in a *package* shall meet the following requirements:

(a) The *dose rate* at 3 m from the unshielded *radioactive material* does not exceed 10 mSv/h.
(b) If subjected to the tests specified in paras 736 and 737, the airborne release in gaseous and particulate forms of up to 100 μm aerodynamic equivalent diameter would not exceed $100A_2$. A separate specimen may be used for each test.
(c) If subjected to the test specified in para. 703, the activity in the water would not exceed $100A_2$. In the application of this test, the damaging effects of the tests specified in subparagraph (b) shall be taken into account.

REQUIREMENTS FOR MATERIAL EXCEPTED FROM FISSILE CLASSIFICATION

606. *Fissile material* excepted from classification as “FISSILE” under para. 417(f) shall be subcritical without the need for accumulation control under the following conditions:

(a) The conditions of para. 673(a);
(b) The conditions consistent with the assessment provisions stated in paras 684(b) and 685(b) for *packages*;
(c) The conditions specified in para. 683(a), if to be transported by air.

GENERAL REQUIREMENTS FOR ALL PACKAGINGS AND PACKAGES

607. The *package* shall be so designed in relation to its mass, volume and shape that it can be easily and safely transported. In addition, the *package* shall be so designed that it can be properly secured in or on the *conveyance* during transport.

608. The *design* shall be such that any lifting attachments on the *package* will not fail when used in the intended manner and that if failure of the attachments should occur, the ability of the *package* to meet other requirements of these Regulations would not be impaired. The *design* shall take account of appropriate safety factors to cover snatch lifting.

609. Attachments and any other features on the outer surface of the *package* that could be used to lift it either shall be designed to support its mass in accordance with the requirements of para. 608 or shall be removable or otherwise rendered incapable of being used during transport.

610. As far as practicable, the *packaging* shall be so designed that the external surfaces are free from protruding features and can be easily decontaminated.

611. As far as practicable, the outer layer of the *package* shall be so designed as to prevent the collection and the retention of water.

612. Any features added to the *package* at the time of transport that are not part of the *package* shall not reduce its safety.

613. The *package* shall be capable of withstanding the effects of any acceleration, vibration or vibration resonance that may arise under routine conditions of transport without any deterioration in the effectiveness of the closing devices on the various receptacles or in the integrity of the *package* as a whole. In particular, nuts, bolts and other securing devices shall be so designed as to prevent them from becoming loose or being released unintentionally, even after repeated use.

613A. The *design* of the *package* shall take into account ageing mechanisms.

614. The materials of the *packaging* and any components or structures shall be physically and chemically compatible with each other and with the *radioactive contents*. Account shall be taken of their behaviour under irradiation.

615. All valves through which the *radioactive contents* could escape shall be protected against unauthorized operation.

616. The *design* of the *package* shall take into account ambient temperatures and pressures that are likely to be encountered in routine conditions of transport.

617. A *package* shall be so designed that it provides sufficient shielding to ensure that, under routine conditions of transport and with the maximum *radioactive contents* that the *package* is designed to contain, the *dose rate* at any point on the external surface of the *package* would not exceed the values specified in para. 516, 527 or 528, as applicable, with account taken of paras 566(b) and 573.

618. For *radioactive material* having other dangerous properties, the *package design* shall take into account those properties (see paras 110 and 507).

ADDITIONAL REQUIREMENTS FOR PACKAGES TO BE TRANSPORTED BY AIR

619. For *packages* to be transported by air, the temperature of the accessible surfaces shall not exceed 50°C at an ambient temperature of 38°C with no account taken for insolation.

620. *Packages* to be transported by air shall be so designed that if they were exposed to ambient temperatures ranging from −40°C to +55°C, the integrity of containment would not be impaired.

621. *Packages* containing *radioactive material* to be transported by air shall be capable of withstanding, without loss or dispersal of *radioactive contents* from the *containment system*, an internal pressure that produces a pressure differential of not less than *maximum normal operating pressure* plus 95 kPa.

REQUIREMENTS FOR EXCEPTED PACKAGES

622. An *excepted package* shall be designed to meet the requirements specified in paras 607–618 and, in addition, the requirements of para. 636 if it contains *fissile material* allowed by one of the provisions of subparagraphs (a)–(h) of para. 417, and the requirements of paras 619–621 if carried by air.

REQUIREMENTS FOR INDUSTRIAL PACKAGES

Requirements for Type IP-1

623. A *Type IP-1 package* shall be designed to meet the requirements specified in paras 607–618 and 636 and, in addition, the requirements of paras 619–621 if carried by air.

Requirements for Type IP-2

624. A *package* to be qualified as *Type IP-2* shall be designed to meet the requirements for *Type IP-1* as specified in para. 623 and, in addition, if it were subjected to the tests specified in paras 722 and 723, it would prevent:

(a) Loss or dispersal of the *radioactive contents*;

(b) More than a 20% increase in the maximum *dose rate* at any external surface of the *package*, except for a *package* whose maximum *dose rate* at its external surface is below 50 μSv/h, for which there shall be no increase of more than 10 μSv/h in the maximum *dose rate* at any external surface of the *package.*

Requirements for Type IP-3

625. A *package* to be qualified as *Type IP-3* shall be designed to meet the requirements for *Type IP-1* as specified in para. 623 and, in addition, the requirements specified in paras 636–649.

Alternative requirements for Type IP-2 and Type IP-3

626. *Packages* may be used as *Type IP-2*, provided that:

(a) They satisfy the requirements for *Type IP-1* specified in para. 623.
(b) They satisfy the requirements prescribed for UN packing group I or II in chapter 6.1 of the United Nations Recommendations on the Transport of Dangerous Goods, Model Regulations (UN Model Regulations) [20].
(c) When subjected to the tests required for UN packing group I or II, they would prevent:
    (i) Loss or dispersal of the *radioactive contents*;
    (ii) More than a 20% increase in the maximum *dose rate* at any external surface of the *package*, except for a *package* whose maximum *dose rate* at its external surface is below 50 μSv/h, for which there shall be no increase of more than 10 μSv/h in the maximum *dose rate* at any external surface of the *package.*

627. *Tanks* may also be used together with their contents as *Type IP-2* or *Type IP-3*, provided that:

(a) They satisfy the requirements for *Type IP-1* specified in para. 623.
(b) They satisfy the requirements prescribed either in:
    (i) Chapter 6.7 of the UN Model Regulations [20], or other requirements, at least equivalent, in the case of portable *tanks*; or
    (ii) Regional or national regulations for the transport of dangerous goods, in the case of *tanks* other than portable *tanks*.
(c) They are capable of withstanding a test pressure of 265 kPa.
(d) Any additional shielding that is provided shall be capable of withstanding the static and dynamic stresses resulting from handling and routine

conditions of transport and of preventing more than a 20% increase in the maximum *dose rate* at any external surface of the *tank*, except for a *tank* whose maximum *dose rate* at its external surface is below 50 μSv/h, for which there shall be no increase of more than 10 μSv/h in the maximum *dose rate* at any external surface of the *tank*.

628. This paragraph has been deleted.

629. *Freight containers* with the characteristics of a permanent enclosure may also be used together with their contents as *Type IP-2* or *Type IP-3*, provided that:

(a) The *radioactive contents* are restricted to solid materials.
(b) They satisfy the requirements for *Type IP-1* specified in para. 623.
(c) They conform to the International Organization for Standardization document: Series 1 Freight Containers — Specification and Testing — Part 1: General Cargo Containers for General Purposes (ISO 1496-1) [21] excluding dimensions and ratings. They shall be so designed that if subjected to the tests prescribed in that document, and to the accelerations occurring during routine conditions of transport, they would prevent:
    (i) Loss or dispersal of the *radioactive contents*;
    (ii) More than a 20% increase in the maximum *dose rate* at any external surface of the *freight container*, except for a *freight container* whose maximum *dose rate* at its external surface is below 50 μSv/h, for which there shall be no increase of more than 10 μSv/h in the maximum *dose rate* at any external surface of the *freight container*.

630. Metal *IBCs* may also be used together with their contents as *Type IP-2* or *Type IP-3*, provided that:

(a) They satisfy the requirements for *Type IP-1* specified in para. 623.
(b) They satisfy the requirements prescribed for UN packing group I or II in chapter 6.5 of the UN Model Regulations [20], and if they were subjected to the tests prescribed in that document, with the drop test conducted in the most damaging orientation, they would prevent:
    (i) Loss or dispersal of the *radioactive contents*;
    (ii) More than a 20% increase in the maximum *dose rate* at any external surface of the *IBC*, except for an *IBC* whose maximum *dose rate* at its external surface is below 50 μSv/h, for which there shall be no increase of more than 10 μSv/h in the maximum *dose rate* at any external surface of the *IBC*.

REQUIREMENTS FOR PACKAGES CONTAINING URANIUM HEXAFLUORIDE

631. *Packages* designed to contain uranium hexafluoride shall meet the requirements that pertain to the radioactive and fissile properties of the material prescribed elsewhere in these Regulations. Except as allowed in para. 634, uranium hexafluoride in quantities of 0.1 kg or more shall also be packaged and transported in accordance with the provisions of ISO 7195 [17], and the requirements of paras 632 and 633.

632. Each *package* designed to contain 0.1 kg or more of uranium hexafluoride shall be so designed that the *package* will meet the following requirements:

(a) Withstand, without leakage and without unacceptable stress, as specified in ISO 7195 [17], the structural test as specified in para. 718, except as allowed in para. 634;
(b) Withstand, without loss or dispersal of the uranium hexafluoride, the free drop test specified in para. 722;
(c) Withstand, without rupture of the *containment system*, the thermal test specified in para. 728, except as allowed in para. 634.

633. *Packages* designed to contain 0.1 kg or more of uranium hexafluoride shall not be provided with pressure relief devices.

634. Subject to *multilateral approval*, *packages* designed to contain 0.1 kg or more of uranium hexafluoride may be transported if the *packages* are designed:

(a) To international or national standards other than ISO 7195 [17], provided an equivalent level of safety is maintained; and/or
(b) To withstand, without leakage and without unacceptable stress, a test pressure of less than 2.76 MPa as specified in para. 718; and/or
(c) To contain 9000 kg or more of uranium hexafluoride and the *packages* do not meet the requirement of para. 632(c).

In all other respects, the requirements specified in paras 631–633 shall be satisfied.

REQUIREMENTS FOR TYPE A PACKAGES

635. *Type A packages* shall be designed to meet the requirements specified in paras 607–618 and, in addition, the requirements of paras 619–621 if carried by air, and of paras 636–651.

636. The smallest overall external dimension of the *package* shall not be less than 10 cm.

637. The outside of the *package* shall incorporate a feature such as a seal that is not readily breakable and that, while intact, will be evidence that the *package* has not been opened.

638. Any tie-down attachments on the *package* shall be so designed that, under normal and accident conditions of transport, the forces in those attachments shall not impair the ability of the *package* to meet the requirements of these Regulations.

639. The *design* of the *package* shall take into account temperatures ranging from −40°C to +70°C for the components of the *packaging*. Attention shall be given to freezing temperatures for liquids and to the potential degradation of *packaging* materials within the given temperature range.

640. The *design* and manufacturing techniques shall be in accordance with national or international standards, or other requirements acceptable to the *competent authority*.

641. The *design* shall include a *containment system* securely closed by a positive fastening device that cannot be opened unintentionally or by a pressure that may arise within the *package*.

642. *Special form radioactive material* may be considered as a component of the *containment system*.

643. If the *containment system* forms a separate unit of the *package*, the *containment system* shall be capable of being securely closed by a positive fastening device that is independent of any other part of the *packaging*.

644. The *design* of any component of the *containment system* shall take into account, where applicable, the radiolytic decomposition of liquids and other vulnerable materials and the generation of gas by chemical reaction and radiolysis.

645. The *containment system* shall retain its *radioactive contents* under a reduction of ambient pressure to 60 kPa.

646. All valves, other than pressure relief valves, shall be provided with an enclosure to retain any leakage from the valve.

647. Radiation shielding that encloses a component of the *package* specified as a part of the *containment system* shall be so designed as to prevent the unintentional release of that component from the shielding. Where the shielding and such a component within it form a separate unit, the shielding shall be capable of being securely closed by a positive fastening device that is independent of any other *packaging* structure.

648. A *package* shall be so designed that if it were subjected to the tests specified in paras 719–724, it would prevent:

(a) Loss or dispersal of the *radioactive contents*;
(b) More than a 20% increase in the maximum *dose rate* at any external surface of the *package*, except for a *package* whose maximum *dose rate* at its external surface is below 50 μSv/h, for which there shall be no increase of more than 10 μSv/h in the maximum *dose rate* at any external surface of the *package*.

649. The *design* of a *package* intended for liquid *radioactive material* shall make provision for ullage to accommodate variations in the temperature of the contents, dynamic effects and filling dynamics.

650. A *Type A package* designed to contain liquid *radioactive material* shall, in addition:

(a) Be adequate to meet the conditions specified in para. 648(a) if the *package* is subjected to the tests specified in para. 725; and
(b) Either:
    (i) Be provided with sufficient absorbent material to absorb twice the volume of the liquid *radioactive contents*. Such absorbent material must be suitably positioned so as to be in contact with the liquid in the event of leakage; or
    (ii) Be provided with a *containment system* composed of primary inner and secondary outer containment components designed to enclose the liquid *radioactive contents* completely and to ensure their retention

within the secondary outer containment components, even if the primary inner components leak.

651. A *Type A package* designed to contain gases shall prevent loss or dispersal of the *radioactive contents* if the *package* were subjected to the tests specified in para. 725, except for a *Type A package* designed to contain tritium gas or noble gases.

REQUIREMENTS FOR TYPE B(U) PACKAGES

652. *Type B(U) packages* shall be designed to meet the requirements specified in paras 607–618, the requirements specified in paras 619–621 if carried by air, and in paras 636–649, except as specified in para. 648(a), and, in addition, the requirements specified in paras 653–666.

653. A *package* shall be so designed that, under an ambient temperature of 38°C and the insolation conditions specified in Table 12 applied prior to the tests, heat generated within the *package* by the *radioactive contents* shall not, under normal conditions of transport, as demonstrated by the tests specified in paras 719–724, adversely affect the *package* in such a way that it would fail to meet the applicable requirements for containment and shielding if left unattended for a period of one week. Particular attention shall be paid to the effects of heat that may cause one or more of the following:

(a) Alteration of the arrangement, the geometrical form or the physical state of the *radioactive contents* or, if the *radioactive material* is enclosed in a can or receptacle (e.g. clad fuel elements), cause the can, receptacle or *radioactive material* to deform or melt;
(b) Lessening of the efficiency of the *packaging* through differential thermal expansion, or cracking or melting of the radiation shielding material;
(c) Acceleration of corrosion when combined with moisture.

654. A *package* shall be so designed that, under an ambient temperature of 38°C and in the absence of insolation, the temperature of the accessible surfaces of a *package* shall not exceed 50°C, unless the *package* is transported under *exclusive use*.

655. Except as required in para. 619 for a *package* to be transported by air, the maximum temperature of any surface readily accessible during transport of a *package* under *exclusive use* shall not exceed 85°C in the absence of insolation

under an ambient temperature of 38°C. Account may be taken of barriers or screens intended to give protection to persons without the need for the barriers or screens being subject to any test.

656. This paragraph has been deleted.

657. The insolation conditions shall be assumed to be as specified in Table 12.

658. A *package* that includes thermal protection for the purpose of satisfying the requirements of the thermal test specified in para. 728 shall be so designed that such protection will remain effective if the *package* is subjected to the tests specified in paras 719–724 and either paras 727(a) and (b), or paras 727(b) and (c), as appropriate. Any such protection on the exterior of the *package* shall not be rendered ineffective by ripping, cutting, skidding, abrading or rough handling.

659. A *package* shall be so designed that if it were subjected to:

(a) The tests specified in paras 719–724, it would restrict the loss of *radioactive contents* to not more than $10^{-6}A_2$ per hour.
(b) The tests specified in paras 726, 727(b), 728 and 729 and either the test in:
  — Para. 727(c), when the *package* has a mass not greater than 500 kg, an overall density not greater than 1000 kg/m^3 based on the external dimensions, and *radioactive contents* greater than $1000A_2$ not as *special form radioactive material*; or
  — Para. 727(a), for all other *packages*.
    (i) It would retain sufficient shielding to ensure that the *dose rate* 1 m from the surface of the *package* would not exceed 10 mSv/h with the maximum *radioactive contents* that the *package* is designed to contain.
    (ii) It would restrict the accumulated loss of *radioactive contents* in a period of one week to not more than 100 TBq for krypton-85 and not more than A_2 for all other radionuclides.

Where mixtures of different radionuclides are present, the provisions of paras 405–407 shall apply, except that for krypton-85 an effective A_2(i) value equal to 100 TBq may be used. For the case in subparagraph (a), the assessment shall take into account the external *non-fixed contamination* limits of para. 508.

660. A *package* for *radioactive contents* with activity greater than 10^5A_2 shall be so designed that if it were subjected to the enhanced water immersion test specified in para. 730, there would be no rupture of the *containment system*.

TABLE 12. INSOLATION CONDITIONS

Case	Form and location of surface	Insolation for 12 h per day (W/m^2)
1	Flat surfaces transported horizontally — downward facing	0
2	Flat surfaces transported horizontally — upward facing	800
3	Surfaces transported vertically	200[a]
4	Other downward facing (not horizontal) surfaces	200[a]
5	All other surfaces	400[a]

[a] Alternatively, a sine function may be used, with an absorption coefficient adopted and the effects of possible reflection from neighbouring objects neglected.

661. Compliance with the permitted activity release limits shall depend neither upon filters nor upon a mechanical cooling system.

662. A *package* shall not include a pressure relief system from the *containment system* that would allow the release of *radioactive material* to the environment under the conditions of the tests specified in paras 719–724 and 726–729.

663. A *package* shall be so designed that if it were at the *maximum normal operating pressure* and it were subjected to the tests specified in paras 719–724 and 726–729, the levels of strain in the *containment system* would not attain values that would adversely affect the *package* in such a way that it would fail to meet the applicable requirements.

664. A *package* shall not have a *maximum normal operating pressure* in excess of a gauge pressure of 700 kPa.

665. A *package* containing *low dispersible radioactive material* shall be so designed that any features added to the *low dispersible radioactive material* that are not part of it, or any internal components of the *packaging*, shall not adversely affect the performance of the *low dispersible radioactive material*.

666. A *package* shall be designed for an ambient temperature range of −40°C to +38°C.

REQUIREMENTS FOR TYPE B(M) PACKAGES

667. *Type B(M) packages* shall meet the requirements for *Type B(U) packages* specified in para. 652, except that:

(a) For a *package* to be transported solely within a specified country or solely between specified countries, ambient temperatures and insolation conditions other than those given in paras 653–655 and 657 and conditions other than those given in paras 639 and 660–666 may be assumed with the *approval* of the *competent authorities* of these countries. The requirements for *Type B(U) packages* specified in paras 655 and 660–666 shall be met as far as practicable.
(b) For a *package* containing only solid *radioactive material* for which the conditions of paras 409(b)(ii) or (c) or the conditions of paras 413(a) or (b) are met, the requirements given in paras 659(a) and (b)(ii) shall not be applicable, provided that, if it were subjected to:
    (i) The tests specified in paras 719–724, it would prevent loss or dispersal of the *radioactive contents*;
    (ii) The tests specified in para. 659(b), any *radioactive material* released from the *package* would be taken into account in demonstration of compliance with the *dose rate* limit in para. 659(b)(i).

668. Intermittent venting of *Type B(M) packages* may be permitted during transport, provided that the operational controls for venting are acceptable to the relevant *competent authorities*.

REQUIREMENTS FOR TYPE C PACKAGES

669. *Type C packages* shall be designed to meet the requirements specified in paras 607–621 and 636–649, except as specified in para. 648(a), and the requirements specified in paras 653–655, 657, 661–666 and 670–672.

670. A *package* shall be capable of meeting the assessment criteria prescribed for tests in paras 659(b) and 663 after burial in an environment with a thermal conductivity of 0.33 W/(m·K) and a temperature of 38°C in the steady state. Initial conditions for the assessment shall assume that any thermal insulation of

the *package* remains intact, the *package* is at the *maximum normal operating pressure* and the ambient temperature is 38°C.

671. A *package* shall be so designed that if it were at the *maximum normal operating pressure* and subjected to:

(a) The tests specified in paras 719–724, it would restrict the loss of *radioactive contents* to not more than $10^{-6}A_2$ per hour.
(b) The test sequences in para. 734:
    (i) It would retain sufficient shielding to ensure that the *dose rate* 1 m from the surface of the *package* would not exceed 10 mSv/h with the maximum *radioactive contents* that the *package* is designed to contain.
    (ii) It would restrict the accumulated loss of *radioactive contents* in a period of one week to not more than 100 TBq for krypton-85 and not more than A_2 for all other radionuclides.

Where mixtures of different radionuclides are present, the provisions of paras 405–407 shall apply, except that for krypton-85 an effective A_2(i) value equal to 100 TBq may be used. For the case in subparagraph (a), the assessment shall take into account the external *non-fixed contamination* limits of para. 508.

672. A *package* shall be so designed that there will be no rupture of the *containment system* following performance of the enhanced water immersion test specified in para. 730.

REQUIREMENTS FOR PACKAGES CONTAINING FISSILE MATERIAL

673. *Fissile material* shall be transported so as to:

(a) Maintain subcriticality during routine, normal and accident conditions of transport. In particular, the following contingencies shall be considered:
    (i) Leakage of water into or out of *packages*;
    (ii) Loss of efficiency of built-in neutron absorbers or moderators;
    (iii) Rearrangement of the contents either within the *package* or as a result of loss from the *package;*
    (iv) Reduction of spaces within or between *packages;*
    (v) *Packages* becoming immersed in water or buried in snow;
    (vi) Temperature changes.

(b) Meet the requirements:
  (i) Of para. 636, except for unpackaged material when specifically allowed by para. 417(e);
  (ii) Prescribed elsewhere in these Regulations that pertain to the radioactive properties of the material;
  (iii) Of para. 637, unless the material is excepted by para. 417;
  (iv) Of paras 676–686, unless the material is excepted by para. 417, 674 or 675.

674. *Packages* containing *fissile material* that meets the requirements of subparagraph (d) and one of the provisions of subparagraphs (a)–(c) of this paragraph are excepted from the requirements of paras 676–686:

(a) *Packages* containing *fissile material* in any form provided that:
  (i) The smallest external dimension of the *package* is not less than 10 cm.
  (ii) The *CSI* of the *package* is calculated using the following formula:
  CSI = 50 × 5 × {[mass of uranium-235 in *package* (g)]/Z
  + [mass of other *fissile nuclides*[1] in *package* (g)]/280}
  where the values of Z are taken from Table 13.
  (iii) The *CSI* of the *package* does not exceed 10.
(b) *Packages* containing *fissile material* in any form provided that:
  (i) The smallest external dimension of the *package* is not less than 30 cm.
  (ii) The *package*, after being subjected to the tests specified in paras 719–724:
    — Retains its *fissile material* contents;
    — Preserves the minimum overall outside dimensions of the *package* to at least 30 cm;
    — Prevents the entry of a 10 cm cube.
  (iii) The *CSI* of the *package* is calculated using the following formula:
  CSI = 50 × 2 × {[mass of uranium-235 in *package* (g)]/Z
  + [mass of other *fissile nuclides*[1] in *package* (g)]/280}
  where the values of Z are taken from Table 13.
  (iv) The *CSI* of the *package* does not exceed 10.

---

[1] Plutonium may be of any isotopic composition provided that the amount of plutonium-241 is less than that of plutonium-240 in the *package*.

TABLE 13. VALUES OF Z FOR CALCULATION OF THE CRITICALITY SAFETY INDEX IN ACCORDANCE WITH PARA. 674

Enrichment[a]	Z
Uranium enriched up to 1.5%	2200
Uranium enriched up to 5%	850
Uranium enriched up to 10%	660
Uranium enriched up to 20%	580
Uranium enriched up to 100%	450

[a] If a *package* contains uranium with varying enrichments of uranium-235, then the value corresponding to the highest enrichment shall be used for Z.

(c) *Packages* containing *fissile material* in any form provided that:
   (i) The smallest external dimension of the *package* is not less than 10 cm.
   (ii) The *package*, after being subjected to the tests specified in paras 719–724:
      — Retains its *fissile material* contents;
      — Preserves the minimum overall outside dimensions of the *package* to at least 10 cm;
      — Prevents the entry of a 10 cm cube.
   (iii) The *CSI* of the *package* is calculated using the following formula:
      CSI = 50 × 2 × {[mass of uranium-235 in *package* (g)]/450
      + [mass of other *fissile nuclides*[1] in *package* (g)]/280}
   (iv) The total mass of *fissile nuclides* in the *package* does not exceed 15 g.

(d) The total mass of beryllium, hydrogenous material enriched in deuterium, graphite and other allotropic forms of carbon in an individual *package* shall not be greater than the mass of *fissile nuclides* in the *package* except where the total concentration of these materials does not exceed 1 g in any 1000 g of material. Beryllium incorporated in copper alloys up to 4% by mass of the alloy does not need to be considered.

675. *Packages* containing not more than 1000 g of plutonium are excepted from the application of paras 676–686 provided that:

(a) Not more than 20% of the plutonium by mass is *fissile nuclides*.
(b) The *CSI* of the *package* is calculated using the following formula:
$CSI = 50 \times 2 \times$ [mass of plutonium (g)/1000]
(c) If uranium is present with the plutonium, the mass of uranium shall be no more than 1% of the mass of the plutonium.

Contents specification for assessments of package designs containing fissile material

676. Where the chemical or physical form, isotopic composition, mass or concentration, moderation ratio or density, or geometric configuration is not known, the assessments of paras 680–685 shall be performed assuming that each parameter that is not known has the value that gives the maximum neutron multiplication consistent with the known conditions and parameters in these assessments.

677. For irradiated nuclear fuel, the assessments in paras 680–685 shall be based on an isotopic composition demonstrated to provide either:

(a) The maximum neutron multiplication during the irradiation history; or
(b) A conservative estimate of the neutron multiplication for the *package* assessments. After irradiation but prior to *shipment*, a measurement shall be performed to confirm the conservatism of the isotopic composition.

Geometry and temperature requirements

678. The *package*, after being subjected to the tests specified in paras 719–724, shall:

(a) Preserve the minimum overall outside dimensions of the *package* to at least 10 cm;
(b) Prevent the entry of a 10 cm cube.

679. The *package* shall be designed for an ambient temperature range of −40°C to +38°C unless the *competent authority* specifies otherwise in the certificate of *approval* for the *package design*.

Assessment of an individual package in isolation

680. For a *package* in isolation, it shall be assumed that water can leak into or out of all void spaces of the *package*, including those within the *containment system*. However, if the *design* incorporates special features to prevent such leakage of water into or out of certain void spaces, even as a result of error, absence of leakage may be assumed in respect of those void spaces. Special features shall include either of the following:

(a) Multiple high standard water barriers, not less than two of which would remain watertight if the *package* were subject to the tests prescribed in para. 685(b), a high degree of quality control in the manufacture, maintenance and repair of *packagings*, and tests to demonstrate the closure of each *package* before each *shipment*; or
(b) For *packages* containing uranium hexafluoride only, with a maximum uranium enrichment of 5 mass percentage of uranium-235:
    (i) *Packages* where, following the tests prescribed in para. 685(b), there is no physical contact between the valve or the plug and any other component of the *packaging* other than at its original point of attachment and where, in addition, following the test prescribed in para. 728, the valve and the plug remain leaktight;
    (ii) A high degree of quality control in the manufacture, maintenance and repair of *packagings*, coupled with tests to demonstrate the closure of each *package* before each *shipment*.

681. It shall be assumed that neutrons that are emitted from the *package* are reflected by at least 20 cm of water.

682. The *package* shall be subcritical under the conditions of paras 680 and 681 and with the *package* conditions that result in the maximum neutron multiplication consistent with:

(a) Routine conditions of transport (incident free);
(b) The tests specified in para. 684(b);
(c) The tests specified in para. 685(b).

683. For *packages* to be transported by air:

(a) The *package* shall be subcritical under conditions consistent with the *Type C package* tests specified in para. 734, assuming that neutrons that are emitted from the *package* are reflected by at least 20 cm of water but that there is no water in-leakage.

(b) In the assessment under para. 682, use of special features as specified in para. 680 is allowed provided that leakage of water into or out of the void spaces is prevented when the *package* is submitted to the *Type C package* tests specified in para. 734 followed by the water leakage test specified in para. 733.

Assessment of package arrays under normal conditions of transport

684. A number N shall be derived, such that five times N *packages* shall be subcritical for the arrangement and *package* conditions that provide the maximum neutron multiplication consistent with the following:

(a) There shall not be anything between the *packages*, and the *package* arrangement shall be reflected on all sides by at least 20 cm of water.
(b) The state of the *packages* shall be their assessed or demonstrated condition if they had been subjected to the tests specified in paras 719–724.

Assessment of package arrays under accident conditions of transport

685. A number N shall be derived, such that two times N *packages* shall be subcritical for the arrangement and *package* conditions that provide the maximum neutron multiplication consistent with the following:

(a) Hydrogenous moderation between the *packages* and the *package* arrangement reflected on all sides by at least 20 cm of water.
(b) The tests specified in paras 719–724 followed by whichever of the following is the more limiting:
    (i) The tests specified in para. 727(b) and in either para. 727(c) for *packages* having a mass not greater than 500 kg and an overall density not greater than 1000 kg/m^3 based on the external dimensions or para. 727(a) for all other *packages*, followed by the test specified in para. 728 and completed by the tests specified in paras 731–733; or
    (ii) The test specified in para. 729.
(c) Where any part of the *fissile material* escapes from the *containment system* following the tests specified in subparagraph (b), it shall be assumed that *fissile material* escapes from each *package* in the array and that all of the *fissile material* shall be arranged in the configuration and moderation that results in the maximum neutron multiplication with close reflection by at least 20 cm of water.

Determination of the criticality safety index for packages

686. The *CSI* for *packages* containing *fissile material* shall be obtained by dividing the number 50 by the smaller of the two values of N derived in paras 684 and 685 (i.e. *CSI* = 50/N). The value of the *CSI* shall not be rounded down except that a value of 0.05 or less for a *package* may be considered as zero.

Section VII

TEST PROCEDURES

DEMONSTRATION OF COMPLIANCE

701. Demonstration of compliance with the performance and acceptance standards prescribed in Section VI shall be accomplished by any of the following methods listed below or by a combination thereof:

(a) Performance of tests with specimens representing *special form radioactive material*, or *low dispersible radioactive material*, or with prototypes or samples of the *packaging*, where the contents of the specimen or the *packaging* for the tests shall simulate as closely as practicable the expected range of *radioactive contents* and the specimen or *packaging* to be tested shall be prepared as presented for transport.

(b) Reference to previous satisfactory demonstrations of a sufficiently similar nature.

(c) Performance of tests with models of appropriate scale, incorporating those features that are significant with respect to the item under investigation when engineering experience has shown the results of such tests to be suitable for *design* purposes. When a scale model is used, the need for adjusting certain test parameters, such as penetrator diameter or compressive load, shall be taken into account.

(d) Calculation, or reasoned argument, when the calculation procedures and parameters are demonstrated to be reliable or conservative.

702. After the specimen, prototype or sample has been subjected to the tests, appropriate methods of assessment shall be used to ensure that the requirements of this section have been fulfilled in compliance with the performance and acceptance standards prescribed in Section VI.

LEACHING TEST FOR LOW DISPERSIBLE RADIOACTIVE MATERIAL

703. A solid material sample representing the entire contents of the *package* shall be immersed for 7 days in water at ambient temperature. The volume of water to be used in the test shall be sufficient to ensure that at the end of the 7 day test period, the free volume of the unabsorbed and unreacted water remaining shall

be at least 10% of the volume of the solid test sample itself. The water shall have an initial pH of 6–8 and a maximum conductivity of 1 mS/m at 20°C. The total activity of the free volume of water shall be measured following the 7 day immersion of the test sample.

TESTS FOR SPECIAL FORM RADIOACTIVE MATERIAL

General

704. Specimens that comprise or simulate *special form radioactive material* shall be subjected to the impact test, the percussion test, the bending test and the heat test specified in paras 705–708. A different specimen may be used for each of the tests. Following each test, a leaching assessment or volumetric leakage test shall be performed on the specimen by a method no less sensitive than the methods given in para. 710 for indispersible solid material or in para. 711 for encapsulated material.

Test methods

705. Impact test: The specimen shall drop onto the target from a height of 9 m. The target shall be as specified in para. 717.

706. Percussion test: The specimen shall be placed on a sheet of lead that is supported by a smooth solid surface and struck by the flat face of a mild steel bar so as to cause an impact equivalent to that resulting from a free drop of 1.4 kg from a height of 1 m. The lower part of the bar shall be 25 mm in diameter with the edges rounded off to a radius of 3.0 ± 0.3 mm. The lead, of hardness number 3.5–4.5 on the Vickers scale and not more than 25 mm thick, shall cover an area greater than that covered by the specimen. A fresh surface of lead shall be used for each impact. The bar shall strike the specimen so as to cause maximum damage.

707. Bending test: The test shall apply only to long, slender sources with both a minimum length of 10 cm and a length to minimum width ratio of not less than 10. The specimen shall be rigidly clamped in a horizontal position so that one half of its length protrudes from the face of the clamp. The orientation of the specimen shall be such that the specimen will suffer maximum damage when its free end is struck by the flat face of a steel bar. The bar shall strike the specimen so as to cause an impact equivalent to that resulting from a free vertical drop of 1.4 kg from a height of 1 m. The lower part of the bar shall be 25 mm in diameter with the edges rounded off to a radius of 3.0 ± 0.3 mm.

708. Heat test: The specimen shall be heated in air to a temperature of 800°C and held at that temperature for a period of 10 min and shall then be allowed to cool.

709. Specimens that comprise or simulate *radioactive material* enclosed in a sealed capsule may be excepted from:

(a) The tests prescribed in paras 705 and 706, provided that the specimens are alternatively subjected to the impact test prescribed in the International Organization for Standardization document: Radiological Protection — Sealed Radioactive Sources — General Requirements and Classification (ISO 2919) [22]:
   (i) The Class 4 impact test if the mass of the *special form radioactive material* is less than 200 g;
   (ii) The Class 5 impact test if the mass of the *special form radioactive material* is more than 200 g but less than 500 g.

(b) The test prescribed in para. 708, provided the specimens are alternatively subjected to the Class 6 temperature test specified in ISO 2919 [22].

Leaching and volumetric leakage assessment methods

710. For specimens that comprise or simulate indispersible solid material, a leaching assessment shall be performed as follows:

(a) The specimen shall be immersed for 7 days in water at ambient temperature. The volume of water to be used in the test shall be sufficient to ensure that at the end of the 7 day test period, the free volume of the unabsorbed and unreacted water remaining shall be at least 10% of the volume of the solid test sample itself. The water shall have an initial pH of 6–8 and a maximum conductivity of 1 mS/m at 20°C.
(b) The water and the specimen shall then be heated to a temperature of 50 ± 5°C and maintained at this temperature for 4 h.
(c) The activity of the water shall then be determined.
(d) The specimen shall then be kept for at least 7 days in still air at not less than 30°C and with a relative humidity of not less than 90%.
(e) The specimen shall then be immersed in water of the same specification as that in subparagraph (a) and the water and the specimen shall be heated to 50 ± 5°C and maintained at this temperature for 4 h.
(f) The activity of the water shall then be determined.

711. For specimens that comprise or simulate *radioactive material* enclosed in a sealed capsule, either a leaching assessment or a volumetric leakage assessment shall be performed as follows:

(a) The leaching assessment shall consist of the following steps:
   (i) The specimen shall be immersed in water at ambient temperature. The water shall have an initial pH of 6–8 with a maximum conductivity of 1 mS/m at 20°C.
   (ii) The water and the specimen shall then be heated to a temperature of 50 ± 5°C and maintained at this temperature for 4 h.
   (iii) The activity of the water shall then be determined.
   (iv) The specimen shall then be kept for at least 7 days in still air at not less than 30°C and with a relative humidity of not less than 90%.
   (v) The process in subparagraphs (i), (ii) and (iii) shall be repeated.
(b) The alternative volumetric leakage assessment shall comprise any of the tests prescribed in ISO 9978 [19] provided that they are acceptable to the *competent authority*.

TESTS FOR LOW DISPERSIBLE RADIOACTIVE MATERIAL

712. A specimen that comprises or simulates *low dispersible radioactive material* shall be subjected to the enhanced thermal test specified in para. 736 and the impact test specified in para. 737. A different specimen may be used for each of the tests. Following each test, the specimen shall be subjected to the leaching test specified in para. 703. After each test it shall be determined if the applicable requirements of para. 605 have been met.

TESTS FOR PACKAGES

Preparation of a specimen for testing

713. All specimens shall be inspected before testing in order to identify and record faults or damage, including the following:

(a) Divergence from the *design*;
(b) Defects in manufacture;
(c) Corrosion or other deterioration;
(d) Distortion of features.

714. The *containment system* of the *package* shall be clearly specified.

715. The external features of the specimen shall be clearly identified so that reference may be made simply and clearly to any part of such a specimen.

Testing the integrity of the containment system and shielding and assessing criticality safety

716. After each test or group of tests or sequence of applicable tests, as appropriate, specified in paras 718–737:

(a) Faults and damage shall be identified and recorded.
(b) It shall be determined whether the integrity of the *containment system* and shielding has been retained to the extent required in Section VI for the *package* under test.
(c) For *packages* containing *fissile material*, it shall be determined whether the assumptions and conditions used in the assessments required by paras 673–686 for one or more *packages* are valid.

Target for drop tests

717. The target for the drop tests specified in paras 705, 722, 725(a), 727 and 735 shall be a flat, horizontal surface of such a character that any increase in its resistance to displacement or deformation upon impact by the specimen would not significantly increase damage to the specimen.

Test for packagings designed to contain uranium hexafluoride

718. Specimens that comprise or simulate *packagings* designed to contain 0.1 kg or more of uranium hexafluoride shall be tested hydraulically at an internal pressure of at least 1.38 MPa, but when the test pressure is less than 2.76 MPa, the *design* shall require *multilateral approval*. For retesting *packagings*, any other equivalent non-destructive testing may be applied, subject to *multilateral approval*.

Tests for demonstrating the ability to withstand normal conditions of transport

719. The tests are the water spray test, the free drop test, the stacking test and the penetration test. Specimens of the *package* shall be subjected to the free drop test, the stacking test and the penetration test, preceded in each case by the

water spray test. One specimen may be used for all the tests, provided that the requirements of para. 720 are fulfilled.

720. The time interval between the conclusion of the water spray test and the succeeding test shall be such that the water has soaked in to the maximum extent, without appreciable drying of the exterior of the specimen. In the absence of any evidence to the contrary, this interval shall be taken to be 2 h if the water spray is applied from four directions simultaneously. No time interval shall elapse, however, if the water spray is applied from each of the four directions consecutively.

721. Water spray test: The specimen shall be subjected to a water spray test that simulates exposure to rainfall of approximately 5 cm per hour for at least 1 h.

722. Free drop test: The specimen shall drop onto the target so as to suffer maximum damage in respect of the safety features to be tested:

(a) The height of the drop, measured from the lowest point of the specimen to the upper surface of the target, shall be not less than the distance specified in Table 14 for the applicable mass. The target shall be as specified in para. 717.
(b) For rectangular fibreboard or wood *packages* not exceeding a mass of 50 kg, a separate specimen shall be subjected to a free drop onto each corner from a height of 0.3 m.
(c) For cylindrical fibreboard *packages* not exceeding a mass of 100 kg, a separate specimen shall be subjected to a free drop onto each of the quarters of each rim from a height of 0.3 m.

TABLE 14. FREE DROP DISTANCE FOR TESTING PACKAGES TO NORMAL CONDITIONS OF TRANSPORT

Package mass (kg)	Free drop distance (m)
package mass < 5 000	1.2
5 000 ≤ *package* mass < 10 000	0.9
10 000 ≤ *package* mass < 15 000	0.6
15 000 ≤ *package* mass	0.3

723. Stacking test: Unless the shape of the *packaging* effectively prevents stacking, the specimen shall be subjected, for a period of 24 h, to a compressive load equal to the greater of the following:

(a) The equivalent of 5 times the maximum mass of the *package*;
(b) The equivalent of 13 kPa multiplied by the vertically projected area of the *package*.

The load shall be applied uniformly to two opposite sides of the specimen, one of which shall be the base on which the *package* would typically rest.

724. Penetration test: The specimen shall be placed on a rigid, flat, horizontal surface that will not move significantly while the test is being carried out:

(a) A bar, 3.2 cm in diameter with a hemispherical end and a mass of 6 kg, shall be dropped and directed to fall with its longitudinal axis vertical onto the centre of the weakest part of the specimen so that, if it penetrates sufficiently far, it will hit the *containment system*. The bar shall not be significantly deformed by the test.
(b) The height of the drop of the bar, measured from its lower end to the intended point of impact on the upper surface of the specimen, shall be 1 m.

Additional tests for Type A packages designed to contain liquids and gases

725. A specimen, or separate specimens, shall be subjected to each of the following tests unless it can be demonstrated that one test is more severe for the specimen in question than the other, in which case one specimen shall be subjected to the more severe test:

(a) Free drop test: The specimen shall drop onto the target so as to suffer the maximum damage in respect of containment. The height of the drop, measured from the lowest part of the specimen to the upper surface of the target, shall be 9 m. The target shall be as specified in para. 717.
(b) Penetration test: The specimen shall be subjected to the test specified in para. 724, except that the height of the drop shall be increased to 1.7 m from the 1 m specified in para. 724(b).

Tests for demonstrating the ability to withstand accident conditions of transport

726. The specimen shall be subjected to the cumulative effects of the tests specified in paras 727 and 728, in that order. Following these tests, either this specimen or a separate specimen shall be subjected to the effect(s) of the water immersion test(s), as specified in para. 729 and, if applicable, para. 730.

727. Mechanical test: The mechanical test consists of three different drop tests. Each specimen shall be subjected to the applicable drops, as specified in para. 659 or 685. The order in which the specimen is subjected to the drops shall be such that, on completion of the mechanical test, the specimen shall have suffered such damage as will lead to maximum damage in the thermal test that follows:

(a) For drop I, the specimen shall drop onto the target so as to suffer maximum damage, and the height of the drop, measured from the lowest point of the specimen to the upper surface of the target, shall be 9 m. The target shall be as specified in para. 717.
(b) For drop II, the specimen shall drop onto a bar rigidly mounted perpendicularly on the target so as to suffer maximum damage. The height of the drop, measured from the intended point of impact of the specimen to the upper surface of the bar, shall be 1 m. The bar shall be of solid mild steel of circular cross-section, 15.0 ± 0.5 cm in diameter and 20 cm long, unless a longer bar would cause greater damage, in which case a bar of sufficient length to cause maximum damage shall be used. The upper end of the bar shall be flat and horizontal with its edge rounded off to a radius of not more than 6 mm. The target on which the bar is mounted shall be as specified in para. 717.
(c) For drop III, the specimen shall be subjected to a dynamic crush test by positioning the specimen on the target so as to suffer maximum damage by the drop of a 500 kg mass from 9 m onto the specimen. The mass shall consist of a solid mild steel plate with dimensions of 1 m × 1 m and shall fall in a horizontal attitude. The lower face of the steel plate shall have its edges and corners rounded off to a radius of not more than 6 mm. The height of the drop shall be measured from the underside of the plate to the highest point of the specimen. The target on which the specimen rests shall be as specified in para. 717.

728. Thermal test: The specimen shall be in thermal equilibrium under conditions of an ambient temperature of 38°C, subject to the insolation conditions specified in Table 12 and subject to the *design* maximum rate of internal heat generation

within the *package* from the *radioactive contents*. Alternatively, any of these parameters are allowed to have different values prior to, and during, the test, provided due account is taken of them in the subsequent assessment of *package* response. The thermal test shall then consist of subparagraph (a) followed by subparagraph (b).

(a) Exposure of a specimen for a period of 30 min to a thermal environment that provides a heat flux at least equivalent to that of a hydrocarbon fuel–air fire in sufficiently quiescent ambient conditions to give a minimum average flame emissivity coefficient of 0.9 and an average temperature of at least 800°C, fully engulfing the specimen, with a surface absorptivity coefficient of 0.8 or that value that the *package* may be demonstrated to possess if exposed to the fire specified.

(b) Exposure of the specimen to an ambient temperature of 38°C, subject to the insolation conditions specified in Table 12 and subject to the *design* maximum rate of internal heat generation within the *package* by the *radioactive contents* for a sufficient period to ensure that temperatures in the specimen are decreasing in all parts of the specimen and/or are approaching initial steady state conditions. Alternatively, any of these parameters are allowed to have different values following cessation of heating, provided due account is taken of them in the subsequent assessment of *package* response. During and following the test, the specimen shall not be artificially cooled and any combustion of materials of the specimen shall be permitted to proceed naturally.

729. Water immersion test: The specimen shall be immersed under a head of water of at least 15 m for a period of not less than 8 h in the attitude that will lead to maximum damage. For demonstration purposes, an external gauge pressure of at least 150 kPa shall be considered to meet these conditions.

Enhanced water immersion test for Type B(U) and Type B(M) packages containing more than 10^5A_2 and for Type C packages

730. Enhanced water immersion test: The specimen shall be immersed under a head of water of at least 200 m for a period of not less than 1 h. For demonstration purposes, an external gauge pressure of at least 2 MPa shall be considered to meet these conditions.

Water leakage test for packages containing fissile material

731. *Packages* for which the water in-leakage or out-leakage that results in greatest reactivity has been assumed for purposes of assessment under paras 680–685 shall be excepted from the water leakage test.

732. Before the specimen is subjected to the water leakage test specified in para. 733, it shall be subjected to the tests specified in para. 727(b) and in either para. 727(a) or (c), as required by para. 685, and the test specified in para. 728.

733. The specimen shall be immersed under a head of water of at least 0.9 m for a period of not less than 8 h and in the attitude for which maximum leakage is expected.

Tests for Type C packages

734. Specimens shall be subjected to the effects of the following test sequences:

(a) The tests specified in paras 727(a), 727(c), 735 and 736, in this order;
(b) The test specified in para. 737.

Separate specimens are allowed to be used for the sequence in subparagraph (a) and for subparagraph (b).

735. Puncture–tearing test: The specimen shall be subjected to the damaging effects of a vertical solid probe made of mild steel. The orientation of the *package* specimen and the impact point on the *package* surface shall be such as to cause maximum damage at the conclusion of the test sequence specified in para. 734(a):

(a) The specimen, representing a *package* having a mass of less than 250 kg, shall be placed on a target and subjected to a probe having a mass of 250 kg falling from a height of 3 m above the intended impact point. For this test the probe shall be a 20 cm diameter cylindrical bar with the striking end forming the frustum of a right circular cone with the following dimensions: 30 cm height and 2.5 cm diameter at the top with its edge rounded off to a radius of not more than 6 mm. The target on which the specimen is placed shall be as specified in para. 717.
(b) For *packages* having a mass of 250 kg or more, the base of the probe shall be placed on a target and the specimen dropped onto the probe. The height of the drop, measured from the point of impact with the specimen to the upper surface of the probe, shall be 3 m. The probe for this test shall have

the same properties and dimensions as specified in subparagraph (a), except that the length and mass of the probe shall be such as to cause maximum damage to the specimen. The target on which the base of the probe is placed shall be as specified in para. 717.

736. Enhanced thermal test: The conditions for this test shall be as specified in para. 728, except that the exposure to the thermal environment shall be for a period of 60 min.

737. Impact test: The specimen shall be subject to an impact on a target at a velocity of not less than 90 m/s, at such an orientation as to suffer maximum damage. The target shall be as specified in para. 717, except that the target surface may be at any orientation as long as the surface is normal to the specimen path.

Section VIII

APPROVAL AND ADMINISTRATIVE REQUIREMENTS

GENERAL

801. For *package designs* where it is not required that a *competent authority* issue a certificate of *approval*, the *consignor* shall, on request, make available for inspection by the relevant *competent authority* documentary evidence of the compliance of the *package design* with all the applicable requirements.

802. *Competent authority approval* shall be required for the following:

(a) *Designs* for:
   (i) *Special form radioactive material* (see paras 803, 804 and 823);
   (ii) *Low dispersible radioactive material* (see paras 803 and 804);
   (iii) *Fissile material* excepted under para. 417(f) (see paras 805 and 806);
   (iv) *Packages* containing 0.1 kg or more of uranium hexafluoride (see para. 807);
   (v) *Packages* containing *fissile material*, unless excepted by para. 417, 674 or 675 (see paras 814–816 and 820);
   (vi) *Type B(U) packages* and *Type B(M) packages* (see paras 808–813 and 820);
   (vii) *Type C packages* (see paras 808–810 and 820).

(b) *Special arrangements* (see paras 829–831).

(c) Certain *shipments* (see paras 825–828).

(d) *Radiation protection programme* for special use *vessels* (see para. 576(a)).

(e) Calculation of radionuclide values that are not listed in the Appendix (see para. 403(a)).

(f) Calculation of alternative activity limits for an exempt *consignment* of instruments or articles (see para. 403(b)).

The certificates of *approval* for the *package design* and the *shipment* may be combined into a single certificate.

APPROVAL OF SPECIAL FORM RADIOACTIVE MATERIAL AND LOW DISPERSIBLE RADIOACTIVE MATERIAL

803. The *design* for *special form radioactive material* shall require *unilateral approval*. The *design* for *low dispersible radioactive material* shall require *multilateral approval*. In both cases, an application for *approval* shall include:

(a) A detailed description of the *radioactive material* or, if a capsule, the contents. Particular reference shall be made to both physical and chemical states.
(b) A detailed statement of the *design* of any capsule to be used.
(c) A statement of the tests that have been performed and their results, or evidence based on calculations, to show that the *radioactive material* is capable of meeting the performance standards, or other evidence that the *special form radioactive material* or *low dispersible radioactive material* meets the applicable requirements of these Regulations.
(d) A specification of the applicable *management system*, as required in para. 306.
(e) Any pre-shipment actions for use in the *consignment* of *special form radioactive material* or *low dispersible radioactive material*.

804. The *competent authority* shall establish a certificate of *approval* stating that the approved *design* meets the requirements for *special form radioactive material* or *low dispersible radioactive material* and shall attribute to that *design* an identification mark.

APPROVAL OF MATERIAL EXCEPTED FROM FISSILE CLASSIFICATION

805. The *design* for *fissile material* excepted from “FISSILE” classification in accordance with Table 1, under para. 417(f) shall require *multilateral approval*. An application for *approval* shall include:

(a) A detailed description of the material. Particular reference shall be made to both the physical and chemical states.
(b) A statement of the tests that have been performed and their results, or evidence based on calculations, to show that the material is capable of meeting the requirements specified in para. 606.

(c) A specification of the applicable *management system,* as required in para. 306.
(d) A statement of specific actions to be taken prior to *shipment.*

806. The *competent authority* shall establish a certificate of *approval* stating that the approved material meets the requirements for *fissile material* excepted by the *competent authority* in accordance with para. 606 and shall attribute to that *design* an identification mark.

APPROVAL OF PACKAGE DESIGNS

Approval of package designs to contain uranium hexafluoride

807. The *approval* of *designs* for *packages* containing 0.1 kg or more of uranium hexafluoride requires that:

(a) Each *design* that meets the requirements of para. 634 shall require *multilateral approval.*
(b) Each *design* that meets the requirements of paras 631–633 shall require *unilateral approval* by the *competent authority* of the country of origin of the *design*, unless *multilateral approval* is otherwise required by these Regulations.
(c) The application for *approval* shall include all information necessary to satisfy the *competent authority* that the *design* meets the requirements of para. 631 and a specification of the applicable *management system*, as required in para. 306.
(d) The *competent authority* shall establish a certificate of *approval* stating that the approved *design* meets the requirements of para. 631 and shall attribute to that *design* an identification mark.

Approval of Type B(U) and Type C package designs

808. Each *Type B(U)* and *Type C package design* shall require *unilateral approval*, except that:

(a) A *package design* for *fissile material*, which is also subject to paras 814–816, shall require *multilateral approval.*
(b) A *Type B(U) package design* for *low dispersible radioactive material* shall require *multilateral approval.*

809. An application for *approval* shall include:

(a) A detailed description of the *radioactive contents* with reference to their physical and chemical states and the nature of the radiation emitted.
(b) A detailed statement of the *design*, including complete engineering drawings and schedules of materials and methods of manufacture.
(c) A statement of the tests that have been performed and their results, or evidence based on calculations or other evidence that the *design* is adequate to meet the applicable requirements.
(d) The operating and maintenance instructions for the use of the *packaging*.
(e) If the *package* is designed to have a *maximum normal operating pressure* in excess of 100 kPa gauge, a specification of the materials of manufacture of the *containment system*, the samples to be taken and the tests to be performed.
(f) A justification of considerations to ageing mechanisms in the safety analysis and within the operating and maintenance instructions.
(g) Where the *radioactive contents* are irradiated nuclear fuel, the applicant shall state and justify any assumption in the safety analysis relating to the characteristics of the fuel and describe any pre-shipment measurement required by para. 677(b).
(h) Any special stowage provisions necessary to ensure the safe dissipation of heat from the *package* considering the various modes of transport to be used and the type of *conveyance* or *freight container*.
(i) A reproducible illustration, not larger than 21 cm × 30 cm, showing the make-up of the *package*.
(j) A specification of the applicable *management system*, as required in para. 306.
(k) For *packages* that are to be used for *shipment* after storage, a gap analysis programme describing a systematic procedure for a periodic evaluation of changes of regulations, changes in technical knowledge and changes of the state of the *package design* during storage.

810. The *competent authority* shall establish a certificate of *approval* stating that the approved *design* meets the requirements for *Type B(U)* or *Type C packages* and shall attribute to that *design* an identification mark.

Approval of Type B(M) package designs

811. Each *Type B(M) package design*, including those for *fissile material*, which are also subject to paras 814–816, and those for *low dispersible radioactive material*, shall require *multilateral approval*.

812. An application for *approval* of a *Type B(M) package design* shall include the information required in para. 809 for *Type B(U) packages* and in addition, when the provisions of para. 667(a) apply, the following:

(a) A list of the requirements specified in paras 639, 653–655, 657 and 660–666 with which the *package* does not conform;
(b) Any proposed supplementary operational controls to be applied during transport that are not provided for in these Regulations but that are necessary to ensure the safety of the *package* or to compensate for the deficiencies listed in subparagraph (a);
(c) A statement relative to any restrictions on the mode of transport and to any special loading, carriage, unloading or handling procedures;
(d) A statement of the range of ambient conditions (e.g. temperature, insolation) that are expected to be encountered during transport and that have been taken into account in the *design*.

813. The *competent authority* shall establish a certificate of *approval* stating that the approved *design* meets the applicable requirements for *Type B(M) packages* and shall attribute to that *design* an identification mark.

Approval of package designs to contain fissile material

814. Each *package design* for *fissile material* that is not excepted by para. 417, 674 or 675 shall require *multilateral approval*.

815. An application for *approval* shall include all information necessary to satisfy the *competent authority* that the *design* meets the requirements of para. 673 and a specification of the applicable *management system*, as required in para. 306.

816. The *competent authority* shall establish a certificate of *approval* stating that the approved *design* meets the requirements of para. 673 and shall attribute to that *design* an identification mark.

APPROVAL OF ALTERNATIVE ACTIVITY LIMITS FOR AN EXEMPT CONSIGNMENT OF INSTRUMENTS OR ARTICLES

817. Alternative activity limits for an exempt *consignment* of instruments or articles in accordance with para. 403(b) shall require *multilateral approval*. An application for *approval* shall include:

(a) An identification, and a detailed description, of the instrument or article, its intended uses and the radionuclide(s) incorporated;
(b) The maximum activity of the radionuclide(s) in the instrument or article;
(c) The maximum external *dose rate* arising from the instrument or article;
(d) The chemical and physical forms of the radionuclide(s) contained in the instrument or article;
(e) Details of the construction and *design* of the instrument or article, particularly as related to the *containment* and shielding of the radionuclide in routine, normal and accident conditions of transport;
(f) The applicable *management system*, including the quality testing and verification procedures to be applied to radioactive sources, components and finished products to ensure that the maximum specified activity of *radioactive material* or the maximum *dose rate* specified for the instrument or article are not exceeded, and that the instruments or articles are constructed in accordance with the *design* specifications;
(g) The maximum number of instruments or articles expected to be shipped per *consignment* and annually;
(h) Dose assessments in accordance with the principles and methodologies set out in GSR Part 3 [2], including individual doses to transport workers and members of the public and, if appropriate, collective doses arising from routine, normal and accident conditions of transport, based on representative transport scenarios that the *consignments* are subject to.

818. The *competent authority* shall establish a certificate of *approval* stating that the approved alternative activity limit for an exempt *consignment* of instruments or articles meets the requirements of para. 403(b) and shall attribute to that certificate an identification mark.

TRANSITIONAL ARRANGEMENTS

Packages not requiring competent authority approval of design under the 1985, 1985 (As Amended 1990), 1996 Edition, 1996 Edition (Revised), 1996 (As Amended 2003), 2005, 2009, 2012 or 2018 Editions of these Regulations

819. *Packages* not requiring *competent authority approval* of *design* (*excepted packages, Type IP-1, Type IP-2, Type IP-3* and *Type A packages*) shall meet this edition of these Regulations in full, except that:

(a) *Packages* that meet the requirements of the 1985 or 1985 (As Amended 1990) Editions of these Regulations:
  (i) May continue in transport provided that they were prepared for transport prior to 31 December 2003 and are subject to the requirements of para. 822, if applicable; or
  (ii) May continue to be used until 31 December 2035, provided that all of the following conditions are met:
    (1) They were not designed to contain uranium hexafluoride.
    (2) The applicable requirements of para. 306 of this edition of these Regulations are applied.
    (3) The activity limits and classification in Section IV of this edition of these Regulations, or of the 2018 Edition of these Regulations, are applied.
    (4) The requirements and controls for transport in Section V of this edition of these Regulations are applied.
    (5) The *packaging* was not manufactured or modified after 31 December 2003.

(b) *Packages* that meet the requirements of the 1996 Edition, 1996 Edition (Revised), 1996 (As Amended 2003), 2005, 2009 or 2012 Editions of these Regulations:
  (i) May continue in transport provided that they were prepared for transport prior to 31 December 2025 and are subject to the requirements of para. 822, if applicable; or
  (ii) May continue to be used, provided that all of the following conditions are met:
    (1) The applicable requirements of para. 306 of this edition of these Regulations are applied.
    (2) The activity limits and classification in Section IV of this edition of these Regulations are applied. Section IV of the 2018 Edition of these Regulations may be used until 31 December 2035.
    (3) The requirements and controls for transport in Section V of this edition of these Regulations are applied.
    (4) The *packaging* was not manufactured or modified after 31 December 2025.

(c) *Packages* that meet the requirements of the 2018 Edition of these Regulations:
  (i) May continue in transport provided that they were prepared for transport prior to 31 December 2035 and are subject to the requirements of para. 822, if applicable; or

(ii) May continue to be used, provided that all of the following conditions are met:
   (1) The activity limits and classification in Section IV of this edition of these Regulations are applied. Section IV of the 2018 Edition of these Regulations may be used until 31 December 2035.
   (2) The requirements and controls for transport in Section V of this edition of these Regulations are applied.
   (3) The *packaging* was not manufactured or modified after 31 December 2035.

Package designs approved under the 1985, 1985 (As Amended 1990), 1996 Edition, 1996 Edition (Revised), 1996 (As Amended 2003), 2005, 2009, 2012 or 2018 Editions of these Regulations

820. *Packages* requiring *competent authority approval* of the *design* shall meet this edition of these Regulations in full except that:

(a) *Packagings* that were manufactured to a *package design* approved by the *competent authority* under the provisions of the 1985 or 1985 (As Amended 1990) Editions of these Regulations may continue to be used provided that all of the following conditions are met:
   (i) The *package design* is subject to *multilateral approval*.
   (ii) The applicable requirements of para. 306 of this edition of these Regulations are applied.
   (iii) The activity limits and classification in Section IV of this edition of these Regulations are applied. Section IV of the 2018 Edition of these Regulations may be used until 31 December 2035.
   (iv) The requirements and controls for transport in Section V of this edition of these Regulations are applied.
   (v) For a *package* containing *fissile material* and transported by air, the requirement of para. 683 is met.
(b) *Packagings* that were manufactured to a *package design* approved by the *competent authority* under the provisions of the 1996 Edition, 1996 Edition (Revised), 1996 (As Amended 2003), 2005, 2009 or 2012 Editions of these Regulations may continue to be used provided that all of the following conditions are met:
   (i) The *package design* is subject to *multilateral approval* after 31 December 2025.
   (ii) The applicable requirements of para. 306 of this edition of the Regulations are applied.

(iii) The activity limits and classification in Section IV of this edition of these Regulations are applied. Section IV of the 2018 Edition of these Regulations may be used until 31 December 2035.

(iv) The requirements and controls for transport in Section V of this edition of these Regulations are applied.

(c) *Packagings* that were manufactured to a *package design* approved by the *competent authority* under the provisions of the 2018 Edition of these Regulations may continue to be used provided that all of the following conditions are met:

(i) The *package design* is subject to *multilateral approval* after 31 December 2035.

(ii) The activity limits and classification in Section IV of this edition of these Regulations are applied. Section IV of the 2018 Edition of these Regulations may be used until 31 December 2035.

(iii) The requirements and controls for transport in Section V of this edition of these Regulations are applied.

821. No new manufacture of *packagings* to a *package design* meeting the provisions of the 1985 or 1985 (As Amended 1990) Editions of these Regulations shall be permitted to commence.

821A. No new manufacture of *packagings* to a *package design* meeting the provisions of the 1996 Edition, 1996 Edition (Revised), 1996 (As Amended 2003), 2005, 2009 or 2012 Editions of these Regulations shall be permitted to commence after 31 December 2028.

821B. No new manufacture of *packagings* to a *package design* meeting the provisions of the 2018 Edition of these Regulations shall be permitted to commence after 31 December 2035.

Packages excepted from the requirements for fissile material under the 2009 Edition of these Regulations

822. *Packages* containing *fissile material* that is excepted from classification as “FISSILE” in accordance with para. 417(a)(i) or (iii) of the 2009 Edition of these Regulations prepared for transport before 31 December 2014 may continue in transport and may continue to be classified as non-fissile or fissile-excepted except that the *consignment* limits in Table 4 of the 2009 Edition of these Regulations shall apply to the *conveyance*. The *conveyance* shall be under *exclusive use*.

Special form radioactive material approved under the 1985, 1985 (As Amended 1990), 1996 Edition, 1996 Edition (Revised), 1996 (As Amended 2003), 2005, 2009, 2012 or 2018 Editions of these Regulations

823. *Special form radioactive material* manufactured to a *design* that had received *unilateral approval* by the *competent authority* under the 1985, 1985 (As Amended 1990), 1996 Edition, 1996 Edition (Revised), 1996 (As Amended 2003), 2005, 2009, 2012 or 2018 Editions of these Regulations may continue to be used when in compliance with the mandatory *management system* in accordance with the applicable requirements of para. 306. There shall be no new manufacture of *special form radioactive material* to a *design* that had received *unilateral approval* by the *competent authority* under the 1985 or 1985 (As Amended 1990) Editions of these Regulations. No new manufacture of *special form radioactive material* to a *design* that had received *unilateral approval* by the *competent authority* under the 1996 Edition, 1996 Edition (Revised), 1996 (As Amended 2003), 2005, 2009 or 2012 Editions of these Regulations shall be permitted to commence after 31 December 2025. No new manufacture of *special form radioactive material* to a *design* that had received *unilateral approval* by the *competent authority* under the 2018 Edition of these Regulations shall be permitted to commence after 31 December 2035.

NOTIFICATION OF SERIAL NUMBERS

824. The *competent authority* shall be notified of the serial number of each *packaging* manufactured to a *design* approved under paras 808, 811, 814 and 820.

APPROVAL OF SHIPMENTS

825. *Multilateral approval* shall be required for:

(a) The *shipment* of *Type B(M) packages* not conforming with the requirements of para. 639 or designed to allow controlled intermittent venting.
(b) The *shipment* of *Type B(M) packages* containing *radioactive material* with an activity greater than 3000A_1 or 3000A_2, as appropriate, or 1000 TBq, whichever is the lower.
(c) The *shipment* of *packages* containing *fissile material* if the sum of the *CSIs* of the *packages* in a single *freight container* or in a single *conveyance* exceeds 50. Excluded from this requirement shall be *shipments* by sea-going *vessels* if the sum of the *CSIs* does not exceed 50 for any hold, compartment

or *defined deck area* and the distance of 6 m between groups of *packages* or *overpacks,* as required in Table 11, is met.

(d) *Radiation protection programmes* for *shipments* by special use *vessels* in accordance with para. 576(a).

(e) The *shipment* of *SCO-III*.

826. A *competent authority* may authorize transport *through or into* its country without *shipment approval*, by a specific provision in its *design approval.*

827. An application for *approval* of *shipment* shall include:

(a) The period of time, related to the *shipment*, for which the *approval* is sought;

(b) The *radioactive contents*, the modes of transport, the type of *conveyance* and the probable or proposed route;

(c) The details of how the precautions and administrative or operational controls, referred to in the certificates of *approval* for the *package design*, if applicable, issued under paras 810, 813 and 816, are to be put into effect.

827A. An application for *approval* of *SCO-III shipments* shall include:

(a) A statement of the respects in which, and of the reasons why, the *consignment* is considered *SCO-III*.

(b) Justification for choosing *SCO-III* by demonstrating that:
  - (i) No suitable *packaging* currently exists;
  - (ii) Designing and/or constructing the *packaging* or segmenting the object is not practically, technically or economically feasible;
  - (iii) No other viable alternative exists.

(c) A detailed description of the *radioactive contents* with reference to their physical and chemical states and the nature of the radiation emitted.

(d) A detailed statement of the *design* of the *SCO-III*, including complete engineering drawings and schedules of materials and methods of manufacture.

(e) All information necessary to satisfy the *competent authority* that the requirements of para. 520(e) and the requirements of para. 522, if applicable, are satisfied.

(f) A transport plan.

(g) A specification of the applicable *management system,* as required in para. 306.

828. Upon *approval* of the *shipment*, the *competent authority* shall issue a certificate of *approval.*

APPROVAL OF SHIPMENTS UNDER SPECIAL ARRANGEMENT

829. Each *consignment* transported under *special arrangement* shall require *multilateral approval.*

830. An application for *approval* of *shipments* under *special arrangement* shall include all the information necessary to satisfy the *competent authority* that the overall level of safety in transport is at least equivalent to that which would be provided if all the applicable requirements of these Regulations had been met. The application shall also include:

(a) A statement of the respects in which, and of the reasons why, the *shipment* cannot be made in full accordance with the applicable requirements;
(b) A statement of any special precautions or special administrative or operational controls that are to be employed during transport to compensate for the failure to meet the applicable requirements.

831. Upon *approval* of *shipments* under *special arrangement*, the *competent authority* shall issue a certificate of *approval.*

COMPETENT AUTHORITY CERTIFICATES OF APPROVAL

Competent authority identification marks

832. Each certificate of *approval* issued by a *competent authority* shall be assigned an identification mark. The mark shall be of the following generalized type:

VRI/Number/Type code

(a) Except as provided in para. 833(b), VRI represents the international *vehicle* registration identification code of the country issuing the certificate.
(b) The number shall be assigned by the *competent authority* and shall be unique and specific with regard to the particular *design, shipment* or alternative activity limit for exempt *consignment.* The identification mark of the *approval* of *shipment* shall be clearly related to the identification mark of the *approval* of the *design.*
(c) The following type codes shall be used in the order listed to indicate the types of certificate of *approval* issued:

AF	*Type A package design* for *fissile material*
B(U)	*Type B(U) package design* (B(U)F if for *fissile material*)

B(M)	*Type B(M) package design* (B(M)F if for *fissile material*)
C	*Type C package design* (CF if for *fissile material*)
IF	*Industrial package design* for *fissile material*
S	*Special form radioactive material*
LD	*Low dispersible radioactive material*
FE	*Fissile material* complying with the requirements of para. 606
T	*Shipment*
X	*Special arrangement*
AL	Alternative activity limits for an exempt *consignment* of instruments or articles

In the case of *package designs* for non-fissile or fissile-excepted uranium hexafluoride or fissile uranium hexafluoride meeting the requirements of para. 674, where none of the above codes apply, the following type codes shall be used:

H(U)	*Unilateral approval*
H(M)	*Multilateral approval*

833. These identification marks shall be applied as follows:

(a) Each certificate and each *package* shall bear the appropriate identification mark comprising the symbols prescribed in para. 832(a)–(c), except that, for *packages*, only the applicable *design* type codes shall appear following the second stroke, that is, the “T” or “X” shall not appear in the identification mark on the *package*. Where the *approval* of the *design* and the *approval* of *shipment* are combined, the applicable type codes do not need to be repeated. For example:

A/132/B(M)F:	A *Type B(M) package design* approved for *fissile material*, requiring *multilateral approval*, for which the *competent authority* of Austria has assigned the *design* number 132 (to be marked both on the *package* and on the certificate of *approval* for the *package design*).
A/132/B(M)FT:	The *approval* of *shipment* issued for a *package* bearing the identification mark elaborated above (to be marked on the certificate only).
A/137/X:	An *approval* of *special arrangement* issued by the *competent authority* of Austria, to which the number 137 has been assigned (to be marked on the certificate only).

A/139/IF: An *industrial package design* for *fissile material* approved by the *competent authority* of Austria, to which *package design* number 139 has been assigned (to be marked both on the *package* and on the certificate of *approval* for the *package design*).

A/145/H(U): A *package design* for fissile-excepted uranium hexafluoride approved by the *competent authority* of Austria, to which *package design* number 145 has been assigned (to be marked both on the *package* and on the certificate of *approval* for the *package design*).

(b) Where *multilateral approval* is effected by validation in accordance with para. 840, only the identification mark issued by the country of origin of the *design* or *shipment* shall be used. Where *multilateral approval* is effected by issue of certificates by successive countries, each certificate shall bear the appropriate identification mark and the *package* whose *design* was so approved shall bear all appropriate identification marks. For example:
A/132/B(M)F
CH/28/B(M)F
would be the identification mark of a *package* that was originally approved by Austria and was subsequently approved, by separate certificate, by Switzerland. Additional identification marks would be tabulated in a similar manner on the *package*.

(c) The revision of a certificate shall be indicated by a parenthetical expression following the identification mark on the certificate.
For example:
A/132/B(M)F (Rev. 2)
would indicate revision 2 of the Austrian certificate of *approval* for the *package design*; or
A/132/B(M)F (Rev. 0)
would indicate the original issuance of the Austrian certificate of *approval* for the *package design*. For original issuances, the parenthetical entry is optional and other words such as “original issuance” may also be used in place of “Rev. 0”. Certificate revision numbers may only be issued by the country issuing the original certificate of *approval*.

(d) Additional symbols (as may be necessitated by national requirements) may be added in brackets to the end of the identification mark.
For example:
A/132/B(M)F (SP503).

(e) It is not necessary to alter the identification mark on the *packaging* each time that a revision to the *design* certificate is made. Such re-marking shall be required only in those cases where the revision to the *package design* certificate involves a change in the letter type codes for the *package design* following the second stroke.

CONTENTS OF CERTIFICATES OF APPROVAL

Certificates of approval for special form radioactive material and low dispersible radioactive material

834. Each certificate of *approval* issued by a *competent authority* for *special form radioactive material* or *low dispersible radioactive material* shall include the following information:

(a) Type of certificate;
(b) The *competent authority* identification mark;
(c) The issue date and an expiry date;
(d) A list of applicable national and international regulations, including the specific edition of these Regulations under which the *special form radioactive material* or *low dispersible radioactive material* is approved;
(e) The identification of the *special form radioactive material* or *low dispersible radioactive material*;
(f) A description of the *special form radioactive material* or *low dispersible radioactive material*;
(g) *Design* specifications for the *special form radioactive material* or *low dispersible radioactive material*, which may include references to drawings;
(h) A specification of the *radioactive contents* that includes the activities involved and which may include the physical and chemical forms;
(i) Reference to the applicable *management system*, as required in para. 306;
(j) Reference to information provided by the applicant relating to specific actions to be taken prior to *shipment*;
(k) If deemed appropriate by the *competent authority*, reference to the identity of the applicant;
(l) Signature and identification of the certifying official.

Certificates of approval for material excepted from fissile classification

835. Each certificate of *approval* issued by a *competent authority* for material excepted from classification as “FISSILE” shall include the following information:

(a) Type of certificate;

(b) The *competent authority* identification mark;
(c) The issue date and an expiry date;
(d) A list of applicable national and international regulations, including the specific edition of these Regulations under which the exception is approved;
(e) A description of the excepted material;
(f) Limiting specifications for the excepted material;
(g) Reference to the applicable *management system*, as required in para. 306;
(h) Reference to information provided by the applicant relating to specific actions to be taken prior to *shipment*;
(i) If deemed appropriate by the *competent authority*, reference to the identity of the applicant;
(j) Signature and identification of the certifying official;
(k) Reference to documentation that demonstrates compliance with para. 606.

Certificates of approval for special arrangement

836. Each certificate of *approval* issued by a *competent authority* for a *special arrangement* shall include the following information:

(a) Type of certificate.
(b) The *competent authority* identification mark.
(c) The issue date and an expiry date.
(d) Mode(s) of transport.
(e) Any restrictions on the modes of transport, type of *conveyance*, *freight container* and any necessary routeing instructions.
(f) A list of applicable national and international regulations, including the specific edition of these Regulations under which the *special arrangement* is approved.
(g) The following statement: "This certificate does not relieve the consignor from compliance with any requirement of the government of any country through or into which the package will be transported".
(h) References to certificates for alternative *radioactive contents*, other *competent authority* validation, or additional technical data or information, as deemed appropriate by the *competent authority*.
(i) A description of the *packaging* by reference to the drawings or a specification of the *design*. If deemed appropriate by the *competent authority*, a reproducible illustration not larger than 21 cm × 30 cm, showing the make-up of the *package*, shall also be provided, accompanied by a brief description of the *packaging*, including materials of manufacture, gross mass, general external dimensions and appearance.

(j) A specification of the authorized *radioactive contents*, including any restrictions on the *radioactive contents* that might not be obvious from the nature of the *packaging*. This specification shall include the physical and chemical forms, the activities involved (including those of the various isotopes, if appropriate), the mass in grams (for *fissile material* or for each *fissile nuclide*, when appropriate) and whether the *special arrangement* is for *special form radioactive material*, *low dispersible radioactive material* or *fissile material* excepted under para. 417(f), if applicable.

(k) Additionally, for *packages* containing *fissile material*:
   - (i) A detailed description of the authorized *radioactive contents*;
   - (ii) The value of the *CSI*;
   - (iii) Reference to the documentation that demonstrates the criticality safety of the *package*;
   - (iv) Any special features on the basis of which the absence of water from certain void spaces has been assumed in the criticality assessment;
   - (v) Any allowance (based on para. 677(b)) for a change in neutron multiplication assumed in the criticality assessment as a result of actual irradiation experience;
   - (vi) The ambient temperature range for which the *special arrangement* has been approved.

(l) A detailed listing of any supplementary operational controls required for preparation, loading, carriage, unloading and handling of the *consignment*, including any special stowage provisions for the safe dissipation of heat.

(m) If deemed appropriate by the *competent authority*, reasons for the *special arrangement*.

(n) A description of the compensatory measures to be applied as a result of the *shipment* being under *special arrangement*.

(o) Reference to information provided by the applicant relating to the use of the *packaging* or specific actions to be taken prior to the *shipment*.

(p) A statement regarding the ambient conditions assumed for purposes of the *design* if these are not in accordance with those specified in paras 653–655, 657 and 666, as applicable.

(q) Any emergency arrangements deemed necessary by the *competent authority*.

(r) Reference to the applicable *management system*, as required in para. 306.

(s) If deemed appropriate by the *competent authority*, reference to the identity of the applicant and to the identity of the *carrier*.

(t) Signature and identification of the certifying official.

Certificates of approval for shipments

837. Each certificate of *approval* for a *shipment* issued by a *competent authority* shall include the following information:

(a) Type of certificate.
(b) The *competent authority* identification mark(s).
(c) The issue date and an expiry date.
(d) A list of applicable national and international regulations, including the specific edition of these Regulations under which the *shipment* is approved.
(e) Any restrictions on the modes of transport, type of *conveyance*, *freight container* and any necessary routeing instructions.
(f) The following statement: "This certificate does not relieve the consignor from compliance with any requirement of the government of any country through or into which the package will be transported".
(g) A detailed listing of any supplementary operational controls required for preparation, loading, carriage, unloading and handling of the *consignment*, including any special stowage provisions for the safe dissipation of heat or maintenance of criticality safety.
(h) Reference to information provided by the applicant relating to specific actions to be taken prior to *shipment*.
(i) Reference to the applicable certificate(s) of *approval* of the *design*.
(j) A specification of the actual *radioactive contents*, including any restrictions on the *radioactive contents* that might not be obvious from the nature of the *packaging*. This specification shall include the physical and chemical forms, the total activities involved (including those of the various isotopes, if appropriate), the mass in grams (for *fissile material* or for each *fissile nuclide*, when appropriate) and whether the *shipment* is for *special form radioactive material*, *low dispersible radioactive material* or *fissile material* excepted under para. 417(f), if applicable.
(k) Any emergency arrangements deemed necessary by the *competent authority*.
(l) Reference to the applicable *management system*, as required in para. 306.
(m) If deemed appropriate by the *competent authority*, reference to the identity of the applicant.
(n) Signature and identification of the certifying official.

Certificates of approval for package design

838. Each certificate of *approval* of the *design* of a *package* issued by a *competent authority* shall include the following information:

(a) Type of certificate.
(b) The *competent authority* identification mark.
(c) The issue date and an expiry date.
(d) Any restriction on the modes of transport, if appropriate.
(e) A list of applicable national and international regulations, including the specific edition of these Regulations under which the *design* is approved.
(f) The following statement: "This certificate does not relieve the consignor from compliance with any requirement of the government of any country through or into which the package will be transported".
(g) References to certificates for alternative *radioactive contents*, other *competent authority* validation, or additional technical data or information, as deemed appropriate by the *competent authority*.
(h) A statement authorizing *shipment*, where *approval* of *shipment* is required under para. 825, if deemed appropriate.
(i) Identification of the *packaging*.
(j) A description of the *packaging* by reference to the drawings or a specification of the *design*. If deemed appropriate by the *competent authority*, a reproducible illustration not larger than 21 cm × 30 cm, showing the make-up of the *package*, shall also be provided, accompanied by a brief description of the *packaging*, including materials of manufacture, gross mass, general external dimensions and appearance.
(k) A specification of the *design* by reference to the drawings.
(l) A specification of the authorized *radioactive contents*, including any restrictions on the *radioactive contents* that might not be obvious from the nature of the *packaging*. This specification shall include the physical and chemical forms, the activities involved (including those of the various isotopes, if appropriate), the mass in grams (for *fissile material*, the total mass of *fissile nuclides* or the mass for each *fissile nuclide*, when appropriate) and whether the *package design* is for *special form radioactive material*, *low dispersible radioactive material* or *fissile material* excepted under para. 417(f), if applicable.
(m) A description of the *containment system*.
(n) For *package designs* containing *fissile material* that require *multilateral approval* of the *package design* in accordance with para. 814:
    (i) A detailed description of the authorized *radioactive contents*;
    (ii) Any design feature of the *package* that is relied on for criticality safety;

   (iii) The value of the *CSI*;
   (iv) Reference to the documentation that demonstrates the criticality safety of the *package*;
   (v) Any special features on the basis of which the absence of water from certain void spaces has been assumed in the criticality assessment;
   (vi) Any allowance (based on para. 677(b)) for a change in neutron multiplication assumed in the criticality assessment as a result of actual irradiation experience;
   (vii) The ambient temperature range for which the *package design* has been approved.

(o) For *Type B(M) packages*, a statement specifying those prescriptions of paras 639, 653–655, 657 and 660–666 with which the *package* does not conform and any amplifying information that may be useful to other *competent authorities*.

(p) For *package designs* subject to para. 820, a statement specifying those requirements of the current regulations with which the *package* does not conform.

(q) For *packages* containing more than 0.1 kg of uranium hexafluoride, a statement specifying those provisions, if any, of para. 634 that apply, and any amplifying information that may be useful to other *competent authorities*.

(r) A detailed listing of any supplementary operational controls required for preparation, loading, carriage, unloading and handling of the *consignment*, including any special stowage provisions for the safe dissipation of heat.

(s) Reference to information provided by the applicant relating to the use of the *packaging* or to specific actions to be taken prior to *shipment*.

(t) A statement regarding the ambient and insolation conditions assumed for purposes of the *design* if these are not in accordance with those specified in paras 653–655, 657 and 666, as applicable.

(u) Reference to the applicable *management system*, as required in para. 306.

(v) Any emergency arrangements deemed necessary by the *competent authority*.

(w) If deemed appropriate by the *competent authority*, reference to the identity of the applicant.

(x) Signature and identification of the certifying official.

Certificates of approval for alternative activity limits for an exempt consignment of instruments or articles

839. Each certificate issued by a *competent authority* for alternative activity limits for an exempt *consignment* of instruments or articles in accordance with para. 818 shall include the following information:

(a) Type of certificate;
(b) The *competent authority* identification mark;
(c) The issue date and an expiry date;
(d) A list of applicable national and international regulations, including the specific edition of these Regulations under which the exemption is approved;
(e) The identification of the instrument or article;
(f) A description of the instrument or article;
(g) *Design* specifications for the instrument or article;
(h) A specification of the radionuclide(s) and the approved alternative activity limit(s) for the exempt *consignment(s)* of the instrument(s) or article(s);
(i) Reference to documentation that demonstrates compliance with para. 403(b);
(j) If deemed appropriate by the *competent authority*, reference to the identity of the applicant;
(k) Signature and identification of the certifying official.

VALIDATION OF CERTIFICATES

840. *Multilateral approval* may be by validation of the original certificate issued by the *competent authority* of the country of origin of the *design* or *shipment*. Such validation may take the form of an endorsement on the original certificate or the issuance of a separate endorsement, annex, supplement, etc., by the *competent authority* of the country *through or into* which the *shipment* is made.

Appendix

BASIC RADIONUCLIDE VALUES

BASIC RADIONUCLIDE VALUES

For footnotes (a)–(k) see pp. 151–154

Radionuclide (atomic number)	A_1 (TBq)	A_2 (TBq)	Activity concentration limit for exempt material (Bq/g)	Activity limit for an exempt *consignment* (Bq)
Actinium (89)				
Ac-225 (a)	4×10^0	7×10^{-4}	1×10^1	1×10^4
Ac-226 (a)	6×10^0	2×10^{-3}	1×10^2	1×10^5
Ac-227 (a)	4×10^1	5×10^{-4}	1×10^{-1}	1×10^3
Ac-228	1×10^0	6×10^{-1}	1×10^1	1×10^6
Silver (47)				
Ag-105	2×10^0	2×10^0	1×10^2	1×10^6
Ag-108m (a)	7×10^{-1}	3×10^{-1}	1×10^1 (b)	1×10^6 (b)
Ag-110m (a)	4×10^{-1}	4×10^{-1}	1×10^1	1×10^6
Ag-111	4×10^1	7×10^{-1}	1×10^3	1×10^6
Aluminium (13)				
Al-26	4×10^{-1}	1×10^{-1}	1×10^1	1×10^5
Americium (95)				
Am-241	4×10^1	2×10^{-3}	1×10^0	1×10^4
Am-242m (a)	4×10^1	2×10^{-3}	1×10^0 (b)	1×10^4 (b)
Am-243 (a)	5×10^0	2×10^{-3}	1×10^0 (b)	1×10^3 (b)

BASIC RADIONUCLIDE VALUES (cont.)

For footnotes (a)–(k) see pp. 151–154

Radionuclide (atomic number)	A_1 (TBq)	A_2 (TBq)	Activity concentration limit for exempt material (Bq/g)	Activity limit for an exempt *consignment* (Bq)
Argon (18)				
Ar-37	Unlimited	Unlimited	1×10^6	1×10^8
Ar-39	4×10^1	2×10^1	1×10^7	1×10^4
Ar-41	9×10^{-1}	3×10^{-1}	1×10^2	1×10^9
Arsenic (33)				
As-72	2×10^{-1}	2×10^{-1}	1×10^1	1×10^5
As-73	4×10^1	3×10^1	1×10^3	1×10^7
As-74	1×10^0	1×10^0	1×10^1	1×10^6
As-76	2×10^{-1}	2×10^{-1}	1×10^2	1×10^5
As-77	4×10^1	7×10^{-1}	1×10^3	1×10^6
Astatine (85)				
At-211 (a)	2×10^1	4×10^{-3}	1×10^3	1×10^7
Gold (79)				
Au-193	8×10^0	3×10^0	1×10^2	1×10^7
Au-194	1×10^0	1×10^0	1×10^1	1×10^6
Au-195	2×10^1	6×10^0	1×10^2	1×10^7
Au-198	3×10^0	7×10^{-1}	1×10^2	1×10^6
Au-199	1×10^1	7×10^{-1}	1×10^2	1×10^6
Barium (56)				
Ba-131 (a)	2×10^0	2×10^0	1×10^2	1×10^6
Ba-133	3×10^0	2×10^0	1×10^2	1×10^6

BASIC RADIONUCLIDE VALUES (cont.)

For footnotes (a)–(k) see pp. 151–154

Radionuclide (atomic number)	A_1 (TBq)	A_2 (TBq)	Activity concentration limit for exempt material (Bq/g)	Activity limit for an exempt *consignment* (Bq)
Ba-133m	2×10^{1}	7×10^{-1}	1×10^{2}	1×10^{6}
Ba-135m	3×10^{1}	6×10^{-1}	1×10^{2}	1×10^{6}
Ba-140 (a)	4×10^{-1}	3×10^{-1}	1×10^{1} (b)	1×10^{5} (b)
Beryllium (4)				
Be-7	2×10^{1}	2×10^{1}	1×10^{3}	1×10^{7}
Be-10	4×10^{1}	6×10^{-1}	1×10^{4}	1×10^{6}
Bismuth (83)				
Bi-205	7×10^{-1}	7×10^{-1}	1×10^{1}	1×10^{6}
Bi-206	3×10^{-1}	3×10^{-1}	1×10^{1}	1×10^{5}
Bi-207	7×10^{-1}	4×10^{-1}	1×10^{1}	1×10^{6}
Bi-210 (a)	4×10^{1}	6×10^{-1}	1×10^{3}	1×10^{6}
Bi-210m (a)	2×10^{0}	2×10^{-3}	1×10^{1}	1×10^{5}
Bi-212 (a)	5×10^{-1}	1×10^{-3}	1×10^{1} (b)	1×10^{5} (b)
Berkelium (97)				
Bk-247	8×10^{0}	2×10^{-3}	1×10^{0}	1×10^{4}
Bk-249 (a)	4×10^{1}	7×10^{-1}	1×10^{3}	1×10^{6}
Bromine (35)				
Br-76	3×10^{-1}	3×10^{-1}	1×10^{1}	1×10^{5}
Br-77	4×10^{0}	4×10^{0}	1×10^{2}	1×10^{6}
Br-82	4×10^{-1}	4×10^{-1}	1×10^{1}	1×10^{6}

BASIC RADIONUCLIDE VALUES (cont.)

For footnotes (a)–(k) see pp. 151–154

Radionuclide (atomic number)	A_1 (TBq)	A_2 (TBq)	Activity concentration limit for exempt material (Bq/g)	Activity limit for an exempt *consignment* (Bq)
Carbon (6)				
C-11	1×10^{0}	6×10^{-1}	1×10^{1}	1×10^{6}
C-14	4×10^{1}	4×10^{0}	1×10^{4}	1×10^{7}
Calcium (20)				
Ca-41	Unlimited	Unlimited	1×10^{5}	1×10^{7}
Ca-45	4×10^{1}	1×10^{0}	1×10^{4}	1×10^{7}
Ca-47 (a)	8×10^{-1}	2×10^{-1}	1×10^{1}	1×10^{6}
Cadmium (48)				
Cd-109	4×10^{1}	3×10^{0}	1×10^{4}	1×10^{6}
Cd-113m	4×10^{1}	8×10^{-1}	1×10^{3}	1×10^{6}
Cd-115 (a)	3×10^{0}	4×10^{-1}	1×10^{2}	1×10^{6}
Cd-115m (a)	1×10^{0}	6×10^{-1}	1×10^{3}	1×10^{6}
Cerium (58)				
Ce-132	5×10^{0}	5×10^{0}	1×10^{1}	1×10^{4}
Ce-133m (a)	5×10^{-1}	5×10^{-1}	1×10^{1}	1×10^{4}
Ce-134 (a)	3×10^{-1}	3×10^{-1}	1×10^{3}	1×10^{7}
Ce-135	1×10^{0}	1×10^{0}	1×10^{1}	1×10^{6}
Ce-137	4×10^{1}	4×10^{1}	1×10^{3}	1×10^{7}
Ce-137m (a)	2×10^{1}	6×10^{-1}	1×10^{3}	1×10^{6}
Ce-139	9×10^{0}	2×10^{0}	1×10^{2}	1×10^{6}
Ce-141	2×10^{1}	7×10^{-1}	1×10^{2}	1×10^{7}

BASIC RADIONUCLIDE VALUES (cont.)

For footnotes (a)–(k) see pp. 151–154

Radionuclide (atomic number)	A_1 (TBq)	A_2 (TBq)	Activity concentration limit for exempt material (Bq/g)	Activity limit for an exempt *consignment* (Bq)
Ce-143	4×10^{0}	6×10^{-1}	1×10^{2}	1×10^{6}
Ce-144 (a)	2×10^{-1}	2×10^{-1}	1×10^{2} (b)	1×10^{5} (b)
Californium (98)				
Cf-248	4×10^{1}	8×10^{-3}	1×10^{1}	1×10^{4}
Cf-249	3×10^{0}	2×10^{-3}	1×10^{0}	1×10^{3}
Cf-250	6×10^{0}	3×10^{-3}	1×10^{1}	1×10^{4}
Cf-251	1×10^{1}	2×10^{-3}	1×10^{0}	1×10^{3}
Cf-252	1×10^{-1}	4×10^{-3}	1×10^{1}	1×10^{4}
Cf-253 (a)	4×10^{1}	1×10^{-1}	1×10^{2}	1×10^{5}
Cf-254	4×10^{-3}	2×10^{-3}	1×10^{0}	1×10^{3}
Chlorine (17)				
Cl-36	4×10^{1}	5×10^{-1}	1×10^{4}	1×10^{6}
Cl-38	2×10^{-1}	2×10^{-1}	1×10^{1}	1×10^{5}
Curium (96)				
Cm-240	4×10^{1}	3×10^{-2}	1×10^{2}	1×10^{5}
Cm-241	2×10^{0}	2×10^{0}	1×10^{2}	1×10^{6}
Cm-242	4×10^{1}	1×10^{-2}	1×10^{2}	1×10^{5}
Cm-243	9×10^{0}	2×10^{-3}	1×10^{0}	1×10^{4}
Cm-244	4×10^{1}	3×10^{-3}	1×10^{1}	1×10^{4}
Cm-245	1×10^{1}	2×10^{-3}	1×10^{0}	1×10^{3}
Cm-246	2×10^{1}	2×10^{-3}	1×10^{0}	1×10^{3}

BASIC RADIONUCLIDE VALUES (cont.)

For footnotes (a)–(k) see pp. 151–154

Radionuclide (atomic number)	A_1 (TBq)	A_2 (TBq)	Activity concentration limit for exempt material (Bq/g)	Activity limit for an exempt *consignment* (Bq)
Cm-247 (a)	3×10^{0}	2×10^{-3}	1×10^{0}	1×10^{4}
Cm-248	7×10^{-2}	5×10^{-4}	1×10^{0}	1×10^{3}
Cobalt (27)				
Co-55	5×10^{-1}	5×10^{-1}	1×10^{1}	1×10^{6}
Co-56	3×10^{-1}	3×10^{-1}	1×10^{1}	1×10^{5}
Co-57	1×10^{1}	1×10^{1}	1×10^{2}	1×10^{6}
Co-58	1×10^{0}	1×10^{0}	1×10^{1}	1×10^{6}
Co-58m	4×10^{1}	4×10^{1}	1×10^{4}	1×10^{7}
Co-60	4×10^{-1}	4×10^{-1}	1×10^{1}	1×10^{5}
Chromium (24)				
Cr-51	4×10^{1}	4×10^{1}	1×10^{3}	1×10^{7}
Caesium (55)				
Cs-129	4×10^{0}	4×10^{0}	1×10^{2}	1×10^{5}
Cs-131	4×10^{1}	4×10^{1}	1×10^{3}	1×10^{6}
Cs-132	2×10^{0}	2×10^{0}	1×10^{1}	1×10^{5}
Cs-134	7×10^{-1}	7×10^{-1}	1×10^{1}	1×10^{4}
Cs-134m	4×10^{1}	8×10^{-1}	1×10^{3}	1×10^{5}
Cs-135	4×10^{1}	1×10^{0}	1×10^{4}	1×10^{7}
Cs-136	5×10^{-1}	5×10^{-1}	1×10^{1}	1×10^{5}
Cs-137 (a)	2×10^{0}	5×10^{-1}	1×10^{1} (b)	1×10^{4} (b)

BASIC RADIONUCLIDE VALUES (cont.)

For footnotes (a)–(k) see pp. 151–154

Radionuclide (atomic number)	A_1 (TBq)	A_2 (TBq)	Activity concentration limit for exempt material (Bq/g)	Activity limit for an exempt *consignment* (Bq)
Copper (29)				
Cu-64	6×10^{0}	1×10^{0}	1×10^{2}	1×10^{6}
Cu-67	1×10^{1}	8×10^{-1}	1×10^{2}	1×10^{6}
Dysprosium (66)				
Dy-159	4×10^{1}	4×10^{1}	1×10^{3}	1×10^{7}
Dy-165	1×10^{1}	7×10^{-1}	1×10^{3}	1×10^{6}
Dy-166 (a)	4×10^{-1}	3×10^{-1}	1×10^{3}	1×10^{6}
Erbium (68)				
Er-169	4×10^{1}	1×10^{0}	1×10^{4}	1×10^{7}
Er-171	3×10^{0}	6×10^{-1}	1×10^{2}	1×10^{6}
Europium (63)				
Eu-147	3×10^{0}	3×10^{0}	1×10^{2}	1×10^{6}
Eu-148	5×10^{-1}	5×10^{-1}	1×10^{1}	1×10^{6}
Eu-149	3×10^{1}	3×10^{1}	1×10^{2}	1×10^{7}
Eu-150	7×10^{-1}	5×10^{-1}	1×10^{1}	1×10^{6}
Eu-150m	2×10^{1}	8×10^{-1}	1×10^{3}	1×10^{6}
Eu-152	1×10^{0}	7×10^{-1}	1×10^{1}	1×10^{6}
Eu-152m	8×10^{-1}	8×10^{-1}	1×10^{2}	1×10^{6}
Eu-154	9×10^{-1}	6×10^{-1}	1×10^{1}	1×10^{6}
Eu-155	2×10^{1}	4×10^{0}	1×10^{2}	1×10^{7}
Eu-156	7×10^{-1}	7×10^{-1}	1×10^{1}	1×10^{6}

BASIC RADIONUCLIDE VALUES (cont.)

For footnotes (a)–(k) see pp. 151–154

Radionuclide (atomic number)	A_1 (TBq)	A_2 (TBq)	Activity concentration limit for exempt material (Bq/g)	Activity limit for an exempt *consignment* (Bq)
Fluorine (9)				
F-18	1×10^{0}	7×10^{-1}	1×10^{1}	1×10^{6}
Iron (26)				
Fe-52 (a)	2×10^{-1}	2×10^{-1}	1×10^{1}	1×10^{6}
Fe-53	2×10^{-1}	2×10^{-1}	1×10^{1}	1×10^{4}
Fe-55	4×10^{1}	4×10^{1}	1×10^{4}	1×10^{6}
Fe-59	9×10^{-1}	9×10^{-1}	1×10^{1}	1×10^{6}
Fe-60 (a)	4×10^{1}	3×10^{-1}	1×10^{2}	1×10^{5}
Gallium (31)				
Ga-67	8×10^{0}	4×10^{0}	1×10^{2}	1×10^{6}
Ga-68	4×10^{-1}	4×10^{-1}	1×10^{1}	1×10^{5}
Ga-72	4×10^{-1}	4×10^{-1}	1×10^{1}	1×10^{5}
Gadolinium (64)				
Gd-146 (a)	4×10^{-1}	4×10^{-1}	1×10^{1}	1×10^{6}
Gd-148	4×10^{1}	4×10^{-3}	1×10^{1}	1×10^{4}
Gd-153	2×10^{1}	1×10^{1}	1×10^{2}	1×10^{7}
Gd-159	2×10^{1}	7×10^{-1}	1×10^{3}	1×10^{6}
Germanium (32)				
Ge-68 (a)	4×10^{-1}	4×10^{-1}	1×10^{1}	1×10^{5}
Ge-69	1×10^{0}	1×10^{0}	1×10^{1}	1×10^{6}
Ge-71	4×10^{1}	4×10^{1}	1×10^{4}	1×10^{8}

BASIC RADIONUCLIDE VALUES (cont.)

For footnotes (a)–(k) see pp. 151–154

Radionuclide (atomic number)	A_1 (TBq)	A_2 (TBq)	Activity concentration limit for exempt material (Bq/g)	Activity limit for an exempt *consignment* (Bq)
Ge-77	6×10^{-1}	6×10^{-1}	1×10^{1}	1×10^{5}
Hydrogen (1)				
H-3 (tritium)	4×10^{1}	4×10^{1}	1×10^{6}	1×10^{9}
Hafnium (72)				
Hf-172 (a)	5×10^{-1}	5×10^{-1}	1×10^{1}	1×10^{6}
Hf-175	3×10^{0}	3×10^{0}	1×10^{2}	1×10^{6}
Hf-181	2×10^{0}	6×10^{-1}	1×10^{1}	1×10^{6}
Hf-182	5×10^{0}	5×10^{0}	1×10^{2}	1×10^{6}
Mercury (80)				
Hg-194 (a)	1×10^{0}	5×10^{-1}	1×10^{1}	1×10^{6}
Hg-195m (a)	3×10^{0}	8×10^{-1}	1×10^{2}	1×10^{6}
Hg-197	2×10^{1}	1×10^{1}	1×10^{2}	1×10^{7}
Hg-197m	1×10^{1}	4×10^{-1}	1×10^{2}	1×10^{6}
Hg-203	5×10^{0}	1×10^{0}	1×10^{2}	1×10^{5}
Holmium (67)				
Ho-166	6×10^{-1}	6×10^{-1}	1×10^{3}	1×10^{5}
Ho-166m	7×10^{-1}	3×10^{-1}	1×10^{1}	1×10^{6}
Iodine (53)				
I-123	8×10^{0}	3×10^{0}	1×10^{2}	1×10^{7}
I-124	1×10^{0}	1×10^{0}	1×10^{1}	1×10^{6}
I-125	4×10^{1}	4×10^{0}	1×10^{3}	1×10^{6}

BASIC RADIONUCLIDE VALUES (cont.)

For footnotes (a)–(k) see pp. 151–154

Radionuclide (atomic number)	A_1 (TBq)	A_2 (TBq)	Activity concentration limit for exempt material (Bq/g)	Activity limit for an exempt *consignment* (Bq)
I-126	3×10^{0}	1×10^{0}	1×10^{2}	1×10^{6}
I-129	4×10^{1}	5×10^{-1}	1×10^{2}	1×10^{5}
I-131	3×10^{0}	8×10^{-1}	1×10^{2}	1×10^{6}
I-132	5×10^{-1}	5×10^{-1}	1×10^{1}	1×10^{5}
I-133	2×10^{0}	7×10^{-1}	1×10^{1}	1×10^{6}
I-134	4×10^{-1}	4×10^{-1}	1×10^{1}	1×10^{5}
I-135 (a)	7×10^{-1}	7×10^{-1}	1×10^{1}	1×10^{6}
Indium (49)				
In-111 (a)	3×10^{0}	3×10^{0}	1×10^{2}	1×10^{6}
In-113m	4×10^{0}	2×10^{0}	1×10^{2}	1×10^{6}
In-114m (a)	5×10^{-1}	3×10^{-1}	1×10^{2}	1×10^{6}
In-115m	7×10^{0}	1×10^{0}	1×10^{2}	1×10^{6}
Iridium (77)				
Ir-189 (a)	2×10^{1}	1×10^{1}	1×10^{2}	1×10^{7}
Ir-190	8×10^{-1}	8×10^{-1}	1×10^{1}	1×10^{6}
Ir-192	1×10^{0} (c)	7×10^{-1}	1×10^{1}	1×10^{4}
Ir-193m	4×10^{1}	5×10^{0}	1×10^{4}	1×10^{7}
Ir-194	4×10^{-1}	4×10^{-1}	1×10^{2}	1×10^{5}
Potassium (19)				
K-40	Unlimited	Unlimited	1×10^{2}	1×10^{6}
K-42	2×10^{-1}	2×10^{-1}	1×10^{2}	1×10^{6}

BASIC RADIONUCLIDE VALUES (cont.)

For footnotes (a)–(k) see pp. 151–154

Radionuclide (atomic number)	A_1 (TBq)	A_2 (TBq)	Activity concentration limit for exempt material (Bq/g)	Activity limit for an exempt *consignment* (Bq)
K-43	1×10^{0}	7×10^{-1}	1×10^{1}	1×10^{6}
Krypton (36)				
Kr-79	4×10^{0}	2×10^{0}	1×10^{3}	1×10^{5}
Kr-81	4×10^{1}	4×10^{1}	1×10^{4}	1×10^{7}
Kr-85	4×10^{1}	2×10^{1}	1×10^{5}	1×10^{4}
Kr-85m	8×10^{0}	3×10^{0}	1×10^{3}	1×10^{10}
Kr-87	2×10^{-1}	2×10^{-1}	1×10^{2}	1×10^{9}
Lanthanum (57)				
La-132	4×10^{-1}	4×10^{-1}	1×10^{1}	1×10^{6}
La-133	8×10^{0}	7×10^{0}	1×10^{1}	1×10^{4}
La-134	3×10^{-1}	3×10^{-1}	1×10^{1}	1×10^{4}
La-135	4×10^{1}	4×10^{1}	1×10^{3}	1×10^{7}
La-137	4×10^{1}	6×10^{0}	1×10^{3}	1×10^{7}
La-140	5×10^{-1}	5×10^{-1}	1×10^{1}	1×10^{5}
Lutetium (71)				
Lu-172	6×10^{-1}	6×10^{-1}	1×10^{1}	1×10^{6}
Lu-173	8×10^{0}	8×10^{0}	1×10^{2}	1×10^{7}
Lu-174	1×10^{1}	9×10^{0}	1×10^{2}	1×10^{7}
Lu-174m	3×10^{1}	9×10^{0}	1×10^{2}	1×10^{7}
Lu-177	4×10^{1}	8×10^{-1}	1×10^{3}	1×10^{7}
Lu-177m	1×10^{0}	4×10^{-1}	1×10^{1}	1×10^{6}

BASIC RADIONUCLIDE VALUES (cont.)

For footnotes (a)–(k) see pp. 151–154

Radionuclide (atomic number)	A_1	A_2	Activity concentration limit for exempt material	Activity limit for an exempt *consignment*
	(TBq)	(TBq)	(Bq/g)	(Bq)
Magnesium (12)				
Mg-28 (a)	2×10^{-1}	2×10^{-1}	1×10^{1}	1×10^{5}
Manganese (25)				
Mn-51	3×10^{-1}	3×10^{-1}	1×10^{1}	1×10^{5}
Mn-52	3×10^{-1}	3×10^{-1}	1×10^{1}	1×10^{5}
Mn-53	Unlimited	Unlimited	1×10^{4}	1×10^{9}
Mn-54	1×10^{0}	1×10^{0}	1×10^{1}	1×10^{6}
Mn-56	3×10^{-1}	3×10^{-1}	1×10^{1}	1×10^{5}
Molybdenum (42)				
Mo-93	4×10^{1}	6×10^{0}	1×10^{3}	1×10^{8}
Mo-99 (a)	4×10^{0}	6×10^{-1}	1×10^{2}	1×10^{6}
Nitrogen (7)				
N-13	1×10^{0}	6×10^{-1}	1×10^{2}	1×10^{9}
Sodium (11)				
Na-22	5×10^{-1}	5×10^{-1}	1×10^{1}	1×10^{6}
Na-24	3×10^{-1}	3×10^{-1}	1×10^{1}	1×10^{5}
Niobium (41)				
Nb-90	3×10^{-1}	3×10^{-1}	1×10^{1}	1×10^{5}
Nb-92m	1×10^{0}	1×10^{0}	1×10^{1}	1×10^{4}
Nb-93m	4×10^{1}	1×10^{1}	1×10^{4}	1×10^{7}
Nb-94	7×10^{-1}	3×10^{-1}	1×10^{1}	1×10^{6}

BASIC RADIONUCLIDE VALUES (cont.)

For footnotes (a)–(k) see pp. 151–154

Radionuclide (atomic number)	A_1	A_2	Activity concentration limit for exempt material	Activity limit for an exempt *consignment*
	(TBq)	(TBq)	(Bq/g)	(Bq)
Nb-95	1×10^{0}	1×10^{0}	1×10^{1}	1×10^{6}
Nb-97	2×10^{0}	6×10^{-1}	1×10^{1}	1×10^{6}
Neodymium (60)				
Nd-147	9×10^{0}	7×10^{-1}	1×10^{2}	1×10^{6}
Nd-149	3×10^{0}	5×10^{-1}	1×10^{2}	1×10^{6}
Nickel (28)				
Ni-56	6×10^{-1}	6×10^{-1}	1×10^{1}	1×10^{4}
Ni-57	6×10^{-1}	6×10^{-1}	1×10^{1}	1×10^{6}
Ni-59	4×10^{1}	4×10^{1}	1×10^{4}	1×10^{8}
Ni-63	4×10^{1}	2×10^{1}	1×10^{5}	1×10^{8}
Ni-65	5×10^{-1}	5×10^{-1}	1×10^{1}	1×10^{6}
Neptunium (93)				
Np-235 (a)	4×10^{1}	4×10^{1}	1×10^{3}	1×10^{7}
Np-236 (a)	9×10^{0}	1×10^{-2}	1×10^{2}	1×10^{5}
Np-236m	3×10^{1}	2×10^{0}	1×10^{3}	1×10^{7}
Np-237	4×10^{1}	2×10^{-3}	1×10^{0} (b)	1×10^{3} (b)
Np-239	7×10^{0}	5×10^{-1}	1×10^{2}	1×10^{7}
Osmium (76)				
Os-185	2×10^{0}	2×10^{0}	1×10^{1}	1×10^{6}
Os-191	2×10^{1}	2×10^{0}	1×10^{2}	1×10^{7}
Os-191m	4×10^{1}	1×10^{1}	1×10^{3}	1×10^{7}

BASIC RADIONUCLIDE VALUES (cont.)

For footnotes (a)–(k) see pp. 151–154

Radionuclide (atomic number)	A_1 (TBq)	A_2 (TBq)	Activity concentration limit for exempt material (Bq/g)	Activity limit for an exempt *consignment* (Bq)
Os-193	2×10^{1}	6×10^{-1}	1×10^{2}	1×10^{6}
Os-194 (a)	4×10^{-1}	4×10^{-1}	1×10^{2}	1×10^{5}
Phosphorus (15)				
P-32	7×10^{-1}	6×10^{-1}	1×10^{3}	1×10^{5}
P-33	4×10^{1}	1×10^{0}	1×10^{5}	1×10^{8}
Protactinium (91)				
Pa-230 (a)	2×10^{0}	2×10^{-1}	1×10^{1}	1×10^{6}
Pa-231	3×10^{1}	5×10^{-4}	1×10^{0}	1×10^{3}
Pa-233	5×10^{0}	7×10^{-1}	1×10^{2}	1×10^{7}
Lead (82)				
Pb-201	1×10^{0}	1×10^{0}	1×10^{1}	1×10^{6}
Pb-202	4×10^{1}	3×10^{-1}	1×10^{3}	1×10^{6}
Pb-203	4×10^{0}	3×10^{0}	1×10^{2}	1×10^{6}
Pb-205	Unlimited	Unlimited	1×10^{4}	1×10^{7}
Pb-210 (a)	4×10^{1}	3×10^{-3}	1×10^{1} (b)	1×10^{4} (b)
Pb-212 (a)	4×10^{-1}	9×10^{-4}	1×10^{1} (b)	1×10^{5} (b)
Palladium (46)				
Pd-103 (a)	4×10^{1}	4×10^{1}	1×10^{3}	1×10^{8}
Pd-107	Unlimited	Unlimited	1×10^{5}	1×10^{8}
Pd-109	4×10^{1}	5×10^{-1}	1×10^{3}	1×10^{6}

BASIC RADIONUCLIDE VALUES (cont.)

For footnotes (a)–(k) see pp. 151–154

Radionuclide (atomic number)	A_1	A_2	Activity concentration limit for exempt material	Activity limit for an exempt *consignment*
	(TBq)	(TBq)	(Bq/g)	(Bq)
Promethium (61)				
Pm-143	4×10^{0}	4×10^{0}	1×10^{2}	1×10^{6}
Pm-144	7×10^{-1}	7×10^{-1}	1×10^{1}	1×10^{6}
Pm-145	4×10^{1}	9×10^{0}	1×10^{3}	1×10^{7}
Pm-147	4×10^{1}	2×10^{0}	1×10^{4}	1×10^{7}
Pm-148m (a)	5×10^{-1}	5×10^{-1}	1×10^{1}	1×10^{6}
Pm-149	4×10^{1}	7×10^{-1}	1×10^{3}	1×10^{6}
Pm-151	4×10^{0}	7×10^{-1}	1×10^{2}	1×10^{6}
Polonium (84)				
Po-210	4×10^{1}	2×10^{-2}	1×10^{1}	1×10^{4}
Praseodymium (59)				
Pr-142	4×10^{-1}	4×10^{-1}	1×10^{2}	1×10^{5}
Pr-143	4×10^{1}	7×10^{-1}	1×10^{4}	1×10^{6}
Platinum (78)				
Pt-188 (a)	4×10^{-1}	4×10^{-1}	1×10^{1}	1×10^{6}
Pt-191	4×10^{0}	3×10^{0}	1×10^{2}	1×10^{6}
Pt-193	4×10^{1}	3×10^{1}	1×10^{4}	1×10^{7}
Pt-193m	4×10^{1}	7×10^{-1}	1×10^{3}	1×10^{7}
Pt-195m	2×10^{1}	6×10^{-1}	1×10^{2}	1×10^{6}
Pt-197	4×10^{1}	7×10^{-1}	1×10^{3}	1×10^{6}
Pt-197m	2×10^{1}	6×10^{-1}	1×10^{2}	1×10^{6}

BASIC RADIONUCLIDE VALUES (cont.)

For footnotes (a)–(k) see pp. 151–154

Radionuclide (atomic number)	A_1 (TBq)	A_2 (TBq)	Activity concentration limit for exempt material (Bq/g)	Activity limit for an exempt *consignment* (Bq)
Plutonium (94)				
Pu-236	4×10^{1}	3×10^{-3}	1×10^{1}	1×10^{4}
Pu-237	3×10^{1}	3×10^{1}	1×10^{3}	1×10^{7}
Pu-238	4×10^{1}	1×10^{-3}	1×10^{0}	1×10^{4}
Pu-239 (a)	4×10^{1}	1×10^{-3}	1×10^{0}	1×10^{4}
Pu-240	4×10^{1}	1×10^{-3}	1×10^{0}	1×10^{3}
Pu-241 (a)	4×10^{1}	6×10^{-2}	1×10^{2}	1×10^{5}
Pu-242	4×10^{1}	1×10^{-3}	1×10^{0}	1×10^{4}
Pu-244 (a)	6×10^{-1}	1×10^{-3}	1×10^{0}	1×10^{4}
Radium (88)				
Ra-223 (a)	2×10^{0}	2×10^{-3}	1×10^{2} (b)	1×10^{5} (b)
Ra-224 (a)	4×10^{-1}	8×10^{-4}	1×10^{1} (b)	1×10^{5} (b)
Ra-225 (a)	1×10^{0}	2×10^{-4}	1×10^{2}	1×10^{5}
Ra-226 (a)	6×10^{-1}	2×10^{-3}	1×10^{1} (b)	1×10^{4} (b)
Ra-228 (a)	1×10^{0}	1×10^{-3}	1×10^{1} (b)	1×10^{5} (b)
Rubidium (37)				
Rb-81 (a)	2×10^{0}	2×10^{0}	1×10^{1}	1×10^{6}
Rb-83 (a)	2×10^{0}	2×10^{0}	1×10^{2}	1×10^{6}
Rb-84	1×10^{0}	1×10^{0}	1×10^{1}	1×10^{6}
Rb-86	6×10^{-1}	6×10^{-1}	1×10^{2}	1×10^{5}
Rb-87	Unlimited	Unlimited	1×10^{4}	1×10^{7}

BASIC RADIONUCLIDE VALUES (cont.)

For footnotes (a)–(k) see pp. 151–154

Radionuclide (atomic number)	A_1	A_2	Activity concentration limit for exempt material	Activity limit for an exempt *consignment*
	(TBq)	(TBq)	(Bq/g)	(Bq)
Rb (natural)	Unlimited	Unlimited	1×10^4	1×10^7
Rhenium (75)				
Re-184	1×10^0	1×10^0	1×10^1	1×10^6
Re-184m	3×10^0	1×10^0	1×10^2	1×10^6
Re-186	4×10^1	7×10^{-1}	1×10^3	1×10^6
Re-187	Unlimited	Unlimited	1×10^6	1×10^9
Re-188	4×10^{-1}	4×10^{-1}	1×10^2	1×10^5
Re-189 (a)	2×10^1	6×10^{-1}	1×10^2	1×10^6
Re (natural)	Unlimited	Unlimited	1×10^6	1×10^9
Rhodium (45)				
Rh-99	2×10^0	2×10^0	1×10^1	1×10^6
Rh-101	4×10^0	4×10^0	1×10^2	1×10^7
Rh-102	2×10^0	2×10^0	1×10^1	1×10^6
Rh-102m	5×10^{-1}	5×10^{-1}	1×10^2	1×10^6
Rh-103m	4×10^1	4×10^1	1×10^4	1×10^8
Rh-105	1×10^1	9×10^{-1}	1×10^2	1×10^7
Radon (86)				
Rn-222 (a)	6×10^{-1}	2×10^{-3}	1×10^1 (b)	1×10^8 (b)
Ruthenium (44)				
Ru-97	5×10^0	5×10^0	1×10^2	1×10^7
Ru-103 (a)	2×10^0	2×10^0	1×10^2	1×10^6

BASIC RADIONUCLIDE VALUES (cont.)

For footnotes (a)–(k) see pp. 151–154

Radionuclide (atomic number)	A_1 (TBq)	A_2 (TBq)	Activity concentration limit for exempt material (Bq/g)	Activity limit for an exempt *consignment* (Bq)
Ru-105	1×10^0	5×10^{-1}	1×10^1	1×10^6
Ru-106 (a)	2×10^{-1}	2×10^{-1}	1×10^2 (b)	1×10^5 (b)
Sulphur (16)				
S-35	4×10^1	4×10^0	1×10^5	1×10^8
Antimony (51)				
Sb-119	4×10^1	4×10^1	1×10^3	1×10^7
Sb-120m	5×10^{-1}	5×10^{-1}	1×10^1	1×10^6
Sb-122	1×10^0	6×10^{-1}	1×10^2	1×10^4
Sb-124	6×10^{-1}	6×10^{-1}	1×10^1	1×10^6
Sb-125	3×10^0	2×10^0	1×10^2	1×10^6
Sb-126	4×10^{-1}	4×10^{-1}	1×10^1	1×10^5
Scandium (21)				
Sc-44	5×10^{-1}	5×10^{-1}	1×10^1	1×10^5
Sc-46	5×10^{-1}	5×10^{-1}	1×10^1	1×10^6
Sc-47	1×10^1	8×10^{-1}	1×10^2	1×10^6
Sc-48	3×10^{-1}	3×10^{-1}	1×10^1	1×10^5
Selenium (34)				
Se-75	3×10^0	3×10^0	1×10^2	1×10^6
Se-79	4×10^1	3×10^0	1×10^4	1×10^7
Silicon (14)				
Si-31	2×10^0	6×10^{-1}	1×10^3	1×10^6

BASIC RADIONUCLIDE VALUES (cont.)

For footnotes (a)–(k) see pp. 151–154

Radionuclide (atomic number)	A_1 (TBq)	A_2 (TBq)	Activity concentration limit for exempt material (Bq/g)	Activity limit for an exempt *consignment* (Bq)
Si-32	4×10^{1}	2×10^{-1}	1×10^{3}	1×10^{6}
Samarium (62)				
Sm-145	4×10^{1}	3×10^{1}	1×10^{2}	1×10^{7}
Sm-147	Unlimited	Unlimited	1×10^{1}	1×10^{4}
Sm-151	4×10^{1}	1×10^{1}	1×10^{4}	1×10^{8}
Sm-153	3×10^{1}	7×10^{-1}	1×10^{2}	1×10^{6}
Tin (50)				
Sn-113 (a)	4×10^{0}	2×10^{0}	1×10^{3}	1×10^{7}
Sn-117m	9×10^{0}	5×10^{-1}	1×10^{2}	1×10^{6}
Sn-119m	4×10^{1}	2×10^{1}	1×10^{3}	1×10^{7}
Sn-121m (a)	4×10^{1}	1×10^{0}	1×10^{3}	1×10^{7}
Sn-123	2×10^{0}	7×10^{-1}	1×10^{3}	1×10^{6}
Sn-125	3×10^{-1}	3×10^{-1}	1×10^{2}	1×10^{5}
Sn-126 (a)	5×10^{-1}	9×10^{-2}	1×10^{1}	1×10^{5}
Strontium (38)				
Sr-82 (a)	2×10^{-1}	2×10^{-1}	1×10^{1}	1×10^{5}
Sr-83	1×10^{0}	1×10^{0}	1×10^{1}	1×10^{6}
Sr-85	2×10^{0}	2×10^{0}	1×10^{2}	1×10^{6}
Sr-85m	5×10^{0}	5×10^{0}	1×10^{2}	1×10^{7}
Sr-87m	3×10^{0}	3×10^{0}	1×10^{2}	1×10^{6}
Sr-89	1×10^{0}	6×10^{-1}	1×10^{3}	1×10^{6}

BASIC RADIONUCLIDE VALUES (cont.)

For footnotes (a)–(k) see pp. 151–154

Radionuclide (atomic number)	A_1	A_2	Activity concentration limit for exempt material	Activity limit for an exempt *consignment*
	(TBq)	(TBq)	(Bq/g)	(Bq)
Sr-90 (a)	3×10^{-1}	1×10^{-1}	1×10^{2} (b)	1×10^{4} (b)
Sr-91 (a)	6×10^{-1}	6×10^{-1}	1×10^{1}	1×10^{5}
Sr-92	8×10^{-1}	8×10^{-1}	1×10^{1}	1×10^{6}
Tritium (1) (see hydrogen)				
Tantalum (73)				
Ta-178 (long lived)	1×10^{0}	9×10^{-1}	1×10^{1}	1×10^{6}
Ta-179	4×10^{1}	4×10^{1}	1×10^{3}	1×10^{7}
Ta-182	9×10^{-1}	6×10^{-1}	1×10^{1}	1×10^{4}
Terbium (65)				
Tb-149	8×10^{-1}	8×10^{-1}	1×10^{1}	1×10^{6}
Tb-157	4×10^{1}	3×10^{1}	1×10^{4}	1×10^{7}
Tb-158	1×10^{0}	5×10^{-1}	1×10^{1}	1×10^{6}
Tb-160	1×10^{0}	6×10^{-1}	1×10^{1}	1×10^{6}
Tb-161	4×10^{1}	8×10^{-1}	1×10^{3}	1×10^{6}
Technetium (43)				
Tc-95	1×10^{0}	1×10^{0}	1×10^{1}	1×10^{6}
Tc-95m (a)	2×10^{0}	2×10^{0}	1×10^{1}	1×10^{6}
Tc-96	4×10^{-1}	4×10^{-1}	1×10^{1}	1×10^{6}
Tc-96m	3×10^{1}	3×10^{1}	1×10^{3}	1×10^{7}
Tc-97	Unlimited	Unlimited	1×10^{3}	1×10^{8}
Tc-97m	4×10^{1}	2×10^{0}	1×10^{3}	1×10^{7}

BASIC RADIONUCLIDE VALUES (cont.)

For footnotes (a)–(k) see pp. 151–154

Radionuclide (atomic number)	A_1 (TBq)	A_2 (TBq)	Activity concentration limit for exempt material (Bq/g)	Activity limit for an exempt *consignment* (Bq)
Tc-98	8×10^{-1}	3×10^{-1}	1×10^{1}	1×10^{6}
Tc-99	4×10^{1}	1×10^{0}	1×10^{4}	1×10^{7}
Tc-99m	1×10^{1}	5×10^{0}	1×10^{2}	1×10^{7}
Tellurium (52)				
Te-118 (a)	2×10^{-1}	2×10^{-1}	1×10^{1}	1×10^{4}
Te-119	1×10^{0}	1×10^{0}	1×10^{1}	1×10^{4}
Te-119m (a)	8×10^{-1}	8×10^{-1}	1×10^{1}	1×10^{4}
Te-121	2×10^{0}	2×10^{0}	1×10^{1}	1×10^{6}
Te-121m	6×10^{0}	3×10^{0}	1×10^{2}	1×10^{6}
Te-123m	9×10^{0}	1×10^{0}	1×10^{2}	1×10^{7}
Te-125m	4×10^{1}	1×10^{0}	1×10^{3}	1×10^{7}
Te-127	4×10^{1}	7×10^{-1}	1×10^{3}	1×10^{6}
Te-127m (a)	4×10^{1}	6×10^{-1}	1×10^{3}	1×10^{7}
Te-129	2×10^{0}	6×10^{-1}	1×10^{2}	1×10^{6}
Te-129m (a)	2×10^{0}	5×10^{-1}	1×10^{3}	1×10^{6}
Te-131m (a)	7×10^{-1}	5×10^{-1}	1×10^{1}	1×10^{6}
Te-132 (a)	4×10^{-1}	4×10^{-1}	1×10^{2}	1×10^{7}
Thorium (90)				
Th-227	9×10^{0}	2×10^{-2}	1×10^{1}	1×10^{4}
Th-228 (a)	5×10^{-1}	9×10^{-4}	1×10^{0} (b)	1×10^{4} (b)
Th-229	1×10^{1}	3×10^{-4}	1×10^{0} (b)	1×10^{3} (b)

BASIC RADIONUCLIDE VALUES (cont.)

For footnotes (a)–(k) see pp. 151–154

Radionuclide (atomic number)	A_1	A_2	Activity concentration limit for exempt material	Activity limit for an exempt *consignment*
	(TBq)	(TBq)	(Bq/g)	(Bq)
Th-230	4×10^{1}	1×10^{-3}	1×10^{0}	1×10^{4}
Th-231	4×10^{1}	1×10^{0}	1×10^{3}	1×10^{7}
Th-232	Unlimited	Unlimited	1×10^{1}	1×10^{4}
Th-234 (a)	4×10^{-1}	4×10^{-1}	1×10^{3} (b)	1×10^{5} (b)
Th (natural)	Unlimited	Unlimited	1×10^{0} (b)	1×10^{3} (b)
Titanium (22)				
Ti-44 (a)	5×10^{-1}	1×10^{-1}	1×10^{1}	1×10^{5}
Thallium (81)				
Tl-200	9×10^{-1}	9×10^{-1}	1×10^{1}	1×10^{6}
Tl-201	2×10^{1}	5×10^{0}	1×10^{2}	1×10^{6}
Tl-202	2×10^{0}	2×10^{0}	1×10^{2}	1×10^{6}
Tl-204	4×10^{1}	8×10^{-1}	1×10^{4}	1×10^{4}
Thulium (69)				
Tm-167	9×10^{0}	9×10^{-1}	1×10^{2}	1×10^{6}
Tm-170	4×10^{1}	7×10^{-1}	1×10^{3}	1×10^{6}
Tm-171	4×10^{1}	4×10^{1}	1×10^{4}	1×10^{8}
Uranium (92)				
U-230 (all lung absorption types) (a)	1×10^{1}	2×10^{-3}	1×10^{1} (b)	1×10^{5} (b)
U-232 (fast lung absorption) (d)	4×10^{1}	3×10^{-2}	1×10^{0} (b)	1×10^{3} (b)

BASIC RADIONUCLIDE VALUES (cont.)

For footnotes (a)–(k) see pp. 151–154

Radionuclide (atomic number)	A_1 (TBq)	A_2 (TBq)	Activity concentration limit for exempt material (Bq/g)	Activity limit for an exempt *consignment* (Bq)
U-232 (between fast and moderate lung absorption) (e)	4×10^1	3×10^{-2}	1×10^0	1×10^3
U-232 (moderate lung absorption) (f)	4×10^1	1×10^{-2}	1×10^0	1×10^3
U-232 (between moderate and slow lung absorption) (g)	4×10^1	1×10^{-3}	1×10^0	1×10^3
U-232 (slow lung absorption) (h)	4×10^1	4×10^{-4}	1×10^0	1×10^3
U-233 (fast lung absorption) (d)	4×10^1	2×10^{-1}	1×10^1	1×10^4
U-233 (between fast and moderate lung absorption) (e)	4×10^1	8×10^{-2}	1×10^1	1×10^4
U-233 (moderate lung absorption) (f)	4×10^1	2×10^{-2}	1×10^1	1×10^4
U-233 (between moderate and slow lung absorption) (g)	4×10^1	6×10^{-3}	1×10^1	1×10^4
U-233 (slow lung absorption) (h)	4×10^1	2×10^{-3}	1×10^1	1×10^4
U-234 (fast lung absorption) (d)	4×10^1	2×10^{-1}	1×10^1	1×10^4
U-234 (between fast and moderate lung absorption) (e)	4×10^1	8×10^{-2}	1×10^1	1×10^4

BASIC RADIONUCLIDE VALUES (cont.)

For footnotes (a)–(k) see pp. 151–154

Radionuclide (atomic number)	A_1	A_2	Activity concentration limit for exempt material	Activity limit for an exempt *consignment*
	(TBq)	(TBq)	(Bq/g)	(Bq)
U-234 (moderate lung absorption) (f)	4×10^{1}	2×10^{-2}	1×10^{1}	1×10^{4}
U-234 (between moderate and slow lung absorption) (g)	4×10^{1}	6×10^{-3}	1×10^{1}	1×10^{4}
U-234 (slow lung absorption) (h)	4×10^{1}	2×10^{-3}	1×10^{1}	1×10^{4}
U-235 (all lung absorption types) (a)	Unlimited	Unlimited	1×10^{1} (b)	1×10^{4} (b)
U-236 (fast lung absorption) (d)	Unlimited	Unlimited	1×10^{1}	1×10^{4}
U-236 (between fast and moderate lung absorption) (e)	Unlimited	Unlimited	1×10^{1}	1×10^{4}
U-236 (moderate lung absorption) (f)	4×10^{1}	2×10^{-2}	1×10^{1}	1×10^{4}
U-236 (between moderate and slow lung absorption) (g)	4×10^{1}	6×10^{-3}	1×10^{1}	1×10^{4}
U-236 (slow lung absorption) (h)	4×10^{1}	2×10^{-3}	1×10^{1}	1×10^{4}
U-238 (all lung absorption types)	Unlimited	Unlimited	1×10^{1} (b)	1×10^{4} (b)
U (natural) (secular equilibrium) (all lung absorption types)	Unlimited	Unlimited	1×10^{0} (b)	1×10^{3} (b)
U (natural) (purified) (all lung absorption types) (i)	Unlimited	Unlimited	1×10^{1} (j)	1×10^{4} (j)

BASIC RADIONUCLIDE VALUES (cont.)

For footnotes (a)–(k) see pp. 151–154

Radionuclide (atomic number)	A_1 (TBq)	A_2 (TBq)	Activity concentration limit for exempt material (Bq/g)	Activity limit for an exempt *consignment* (Bq)
U (enriched to 10% or less) (all lung absorption types) (k)	Unlimited	Unlimited	1×10^{1} (j)	1×10^{4} (j)
U (enriched to 20% or less) (except slow lung absorption) (d)(e)(f)(g)(k)	Unlimited	Unlimited	1×10^{1} (j)	1×10^{4} (j)
U (enriched to 20% or less) (slow lung absorption) (h)(k)	1×10^{1}	2×10^{-3}	1×10^{1} (j)	1×10^{4} (j)
U (depleted) (all lung absorption types)	Unlimited	Unlimited	1×10^{1} (j)	1×10^{4} (j)
Vanadium (23)				
V-48	4×10^{-1}	4×10^{-1}	1×10^{1}	1×10^{5}
V-49	4×10^{1}	4×10^{1}	1×10^{4}	1×10^{7}
Tungsten (74)				
W-178 (a)	1×10^{1}	5×10^{0}	1×10^{1}	1×10^{6}
W-181	4×10^{1}	4×10^{1}	1×10^{3}	1×10^{7}
W-185	4×10^{1}	9×10^{-1}	1×10^{4}	1×10^{7}
W-187	2×10^{0}	7×10^{-1}	1×10^{2}	1×10^{6}
W-188 (a)	4×10^{-1}	4×10^{-1}	1×10^{2}	1×10^{5}
Xenon (54)				
Xe-122 (a)	2×10^{-1}	2×10^{-1}	1×10^{2}	1×10^{9}
Xe-123	2×10^{0}	7×10^{-1}	1×10^{2}	1×10^{9}
Xe-127	5×10^{0}	2×10^{0}	1×10^{3}	1×10^{5}

BASIC RADIONUCLIDE VALUES (cont.)

For footnotes (a)–(k) see pp. 151–154

Radionuclide (atomic number)	A_1 (TBq)	A_2 (TBq)	Activity concentration limit for exempt material (Bq/g)	Activity limit for an exempt *consignment* (Bq)
Xe-131m	4×10^{1}	4×10^{1}	1×10^{4}	1×10^{4}
Xe-133	4×10^{1}	2×10^{1}	1×10^{3}	1×10^{4}
Xe-135	5×10^{0}	2×10^{0}	1×10^{3}	1×10^{10}
Yttrium (39)				
Y-87 (a)	1×10^{0}	1×10^{0}	1×10^{1}	1×10^{6}
Y-88	4×10^{-1}	4×10^{-1}	1×10^{1}	1×10^{6}
Y-89m	1×10^{0}	1×10^{0}	1×10^{1}	1×10^{4}
Y-90	3×10^{-1}	3×10^{-1}	1×10^{3}	1×10^{5}
Y-91	1×10^{0}	6×10^{-1}	1×10^{3}	1×10^{6}
Y-91m	2×10^{0}	2×10^{0}	1×10^{2}	1×10^{6}
Y-92	2×10^{-1}	2×10^{-1}	1×10^{2}	1×10^{5}
Y-93	2×10^{-1}	2×10^{-1}	1×10^{2}	1×10^{5}
Ytterbium (70)				
Yb-169	4×10^{0}	1×10^{0}	1×10^{2}	1×10^{7}
Yb-175	3×10^{1}	1×10^{0}	1×10^{3}	1×10^{7}
Zinc (30)				
Zn-65	2×10^{0}	2×10^{0}	1×10^{1}	1×10^{6}
Zn-69	4×10^{1}	7×10^{-1}	1×10^{4}	1×10^{6}
Zn-69m (a)	3×10^{0}	6×10^{-1}	1×10^{2}	1×10^{6}
Zirconium (40)				
Zr-88	3×10^{0}	3×10^{0}	1×10^{2}	1×10^{6}

BASIC RADIONUCLIDE VALUES (cont.)

For footnotes (a)–(k) see pp. 151–154

Radionuclide (atomic number)	A_1	A_2	Activity concentration limit for exempt material	Activity limit for an exempt *consignment*
	(TBq)	(TBq)	(Bq/g)	(Bq)
Zr-89	9×10^{-1}	9×10^{-1}	1×10^{1}	1×10^{6}
Zr-93	Unlimited	Unlimited	1×10^{3} (b)	1×10^{7} (b)
Zr-95 (a)	1×10^{0}	1×10^{0}	1×10^{1}	1×10^{6}
Zr-97 (a)	4×10^{-1}	3×10^{-1}	1×10^{1} (b)	1×10^{5} (b)

(a) A_1 and/or A_2 values for these parent radionuclides include contributions from their progeny with half-lives less than 10 days, as listed in the following:

Mg-28	Al-28
Ca-47	Sc-47
Ti-44	Sc-44
Fe-52	Mn-52m
Fe-60	Co-60m
Zn-69m	Zn-69
Ge-68	Ga-68
Rb-81	Kr-81m
Rb-83	Kr-83m
Sr-82	Rb-82
Sr-90	Y-90
Sr-91	Y-91m
Y-87	Sr-87m
Zr-95	Nb-95m
Zr-97	Nb-97
Mo-99	Tc-99m
Tc-95m	Tc-95
Ru-103	Rh-103m
Ru-106	Rh-106
Pd-103	Rh-103m
Ag-108m	Ag-108
Ag-110m	Ag-110
Cd-115	In-115m
Cd-115m	In-115m
In-111	Cd-111m

Footnote (a) (cont.)

In-114m	In-114
Sn-113	In-113m
Sn-121m	Sn-121
Sn-126	Sb-126m
Te-118	Sb-118
Te-119m	Sb-119
Te-127m	Te-127
Te-129m	Te-129
Te-131m	Te-131
Te-132	I-132
Xe-122	I-122
Ba-131	Cs-131
Ce-133m	La-133
Ce-134	La-134
Ce-137m	Ce-137
Ce-144	Pr-144m, Pr-144
I-135	Xe-135m
Cs-137	Ba-137m
Ba-140	La-140
Pm-148m	Pm-148
Gd-146	Eu-146
Dy-166	Ho-166
Hf-172	Lu-172, Lu-172m
W-178	Ta-178
W-188	Re-188
Re-189	Os-189m
Ir-189	Os-189m
Pt-188	Ir-188
Os-194	Ir-194
Hg-194	Au-194
Hg-195m	Hg-195
Pb-210	Bi-210, Hg-206, Tl-206
Pb-212	Bi-212, Tl-208, Po-212
Bi-210	Tl-206
Bi-210m	Tl-206
Bi-212	Tl-208, Po-212
At-211	Po-211
Rn-222	Po-218, Pb-214, At-218, Bi-214, Rn-218, Po-214, Tl-210
Ra-223	Rn-219, Po-215, Pb-211, Bi-211, Po-211, Tl-207
Ra-224	Rn-220, Po-216, Pb-212, Bi-212, Tl-208, Po-212
Ra-225	Ac-225, Fr-221, At-217, Bi-213, Tl-209, Po-213, Pb-209
Ra-226	Rn-222, Po-218, Pb-214, At-218, Bi-214, Rn-218, Po-214, Tl-210
Ra-228	Ac-228
Ac-225	Fr-221, At-217, Bi-213, Tl-209, Po-213, Pb-209

Footnote (a) (cont.)

Ac-226	Th-226, Fr-222, Ra-222, Rn-218, Po-214
Ac-227	Fr-223, At-219, Bi-215, Po-215, Pb-211, Bi-211, Tl-207, Po-211
Th-228	Ra-224, Rn-220, Po-216, Pb-212, Bi-212, Tl-208, Po-212
Th-234	Pa-234m, Pa-234
Pa-230	Ac-226, Th-226, Fr-222, Ra-222, Rn-218, Po-214
U-230	Th-226, Ra-222, Rn-218, Po-214
U-235	Th-231
Np-235	U-235m
Np-236	Pa-232
Pu-239	U-235m
Pu-241	U-237
Pu-244	U-240, Np-240m, Np-240
Am-242m	Am-242, Np-238
Am-243	Np-239
Cm-247	Pu-243
Bk-249	Am-245
Cf-253	Cm-249

(b) Parent nuclides and their progeny included in secular equilibrium are listed in the following (the activity to be taken into account is that of the parent nuclide only):

Sr-90	Y-90
Zr-93	Nb-93m
Zr-97	Nb-97
Ru-106	Rh-106
Ag-108m	Ag-108
Cs-137	Ba-137m
Ce-144	Pr-144
Ba-140	La-140
Bi-212	Tl-208 (0.36), Po-212 (0.64)
Pb-210	Bi-210, Po-210
Pb-212	Bi-212, Tl-208 (0.36), Po-212 (0.64)
Rn-222	Po-218, Pb-214, Bi-214, Po-214
Ra-223	Rn-219, Po-215, Pb-211, Bi-211, Tl-207
Ra-224	Rn-220, Po-216, Pb-212, Bi-212, Tl-208 (0.36), Po-212 (0.64)
Ra-226	Rn-222, Po-218, Pb-214, Bi-214, Po-214, Pb-210, Bi-210, Po-210
Ra-228	Ac-228
Th-228	Ra-224, Rn-220, Po-216, Pb-212, Bi-212, Tl-208 (0.36), Po-212 (0.64)
Th-229	Ra-225, Ac-225, Fr-221, At-217, Bi-213, Po-213, Pb-209
Th-234	Pa-234m
Th (natural)*	Ra-228, Ac-228, Th-228, Ra-224, Rn-220, Po-216, Pb-212, Bi-212, Tl-208 (0.36), Po-212 (0.64)
U-230	Th-226, Ra-222, Rn-218, Po-214

Footnote (b) (cont.)

U-232	Th-228, Ra-224, Rn-220, Po-216, Pb-212, Bi-212, Tl-208 (0.36), Po-212 (0.64)
U-235	Th-231
U-238	Th-234, Pa-234m
U (natural)*	Th-234, Pa-234m, U-234, Th-230, Ra-226, Rn-222, Po-218, Pb-214, Bi-214, Po-214, Pb-210, Bi-210, Po-210
Np-237	Pa-233
Am-242m	Am-242
Am-243	Np-239

* In the case of Th (natural), the parent nuclide is Th-232; in the case of U (natural), the parent nuclide is U-238.

(c) The quantity may be determined from a measurement of the rate of decay or a measurement of the *dose rate* at a prescribed distance from the source.

(d) These values apply only to chemical forms of uranium with a fast lung absorption rate, including UF_6, UO_2F_2 and uranyl tributyl phosphate in both normal and accident conditions of transport.

(e) These values apply only to chemical forms of uranium with lung absorption rates between fast and moderate, including uranyl nitrate ($UO_2(NO_3)_2$), UO_4, ammonium diuranate and UO_3 in both normal and accident conditions of transport.

(f) These values apply only to chemical forms of uranium with a moderate lung absorption rate, including uranyl acetylacetonate, UF_4, UCl_4, hexavalent compounds, *depleted uranium* aerosols from the use of kinetic energy penetrators, and vaporized uranium metal in both normal and accident conditions of transport.

(g) These values apply only to chemical forms of uranium with lung absorption rates between moderate and slow, including U_3O_8, UO_2, uranium aluminide (UAl_x), and all chemical forms of uranium other than those specified in footnotes (d), (e), (f) above and (h) below in both normal and accident conditions of transport.

(h) These values apply to chemical forms of uranium with a slow lung absorption rate.

(i) These values apply only to *natural uranium* that has undergone chemical purification after mining.

(j) Parent nuclides and their progenies included in secular equilibrium are listed in the following (the activity to be taken into account is the sum of the activities of U-234, U-235 and U-238 only):

U-234	
U-235	Th-231
U-238	Th-234, Pa-234m

(k) These values apply to *unirradiated uranium* only.

REFERENCES

References are to editions that are current as of the time of publication of these Regulations. Editions that supersede these may be adopted under national legislation.

[1] EUROPEAN ATOMIC ENERGY COMMUNITY, FOOD AND AGRICULTURE ORGANIZATION OF THE UNITED NATIONS, INTERNATIONAL ATOMIC ENERGY AGENCY, INTERNATIONAL LABOUR ORGANIZATION, INTERNATIONAL MARITIME ORGANIZATION, OECD NUCLEAR ENERGY AGENCY, PAN AMERICAN HEALTH ORGANIZATION, UNITED NATIONS ENVIRONMENT PROGRAMME, WORLD HEALTH ORGANIZATION, Fundamental Safety Principles, IAEA Safety Standards Series No. SF-1, IAEA, Vienna (2006), https://doi.org/10.61092/iaea.hmxn-vw0a

[2] EUROPEAN COMMISSION, FOOD AND AGRICULTURE ORGANIZATION OF THE UNITED NATIONS, INTERNATIONAL ATOMIC ENERGY AGENCY, INTERNATIONAL LABOUR ORGANIZATION, OECD NUCLEAR ENERGY AGENCY, PAN AMERICAN HEALTH ORGANIZATION, UNITED NATIONS ENVIRONMENT PROGRAMME, WORLD HEALTH ORGANIZATION, Radiation Protection and Safety of Radiation Sources: International Basic Safety Standards, IAEA Safety Standards Series No. GSR Part 3, IAEA, Vienna (2014), https://doi.org/10.61092/iaea.u2pu-60vm

[3] INTERNATIONAL ATOMIC ENERGY AGENCY, Governmental, Legal and Regulatory Framework for Safety, IAEA Safety Standards Series No. GSR Part 1 (Rev. 1), IAEA, Vienna (2016).

[4] INTERNATIONAL ATOMIC ENERGY AGENCY, Leadership and Management for Safety, IAEA Safety Standards Series No. GSR Part 2, IAEA, Vienna (2016), https://doi.org/10.61092/iaea.cq1k-j5z3

[5] INTERNATIONAL ATOMIC ENERGY AGENCY, Safety Assessment for Facilities and Activities, IAEA Safety Standards Series No. GSR Part 4 (Rev. 1), IAEA, Vienna (2016).

[6] FOOD AND AGRICULTURE ORGANIZATION OF THE UNITED NATIONS, INTERNATIONAL ATOMIC ENERGY AGENCY, INTERNATIONAL CIVIL AVIATION ORGANIZATION, INTERNATIONAL LABOUR ORGANIZATION, INTERNATIONAL MARITIME ORGANIZATION, INTERPOL, OECD NUCLEAR ENERGY AGENCY, PAN AMERICAN HEALTH ORGANIZATION, PREPARATORY COMMISSION FOR THE COMPREHENSIVE NUCLEAR-TEST-BAN TREATY ORGANIZATION, UNITED NATIONS ENVIRONMENT PROGRAMME, UNITED NATIONS OFFICE FOR THE COORDINATION OF HUMANITARIAN AFFAIRS, WORLD HEALTH ORGANIZATION, WORLD METEOROLOGICAL ORGANIZATION, Preparedness and Response for a Nuclear or Radiological Emergency, IAEA Safety Standards Series No. GSR Part 7, IAEA, Vienna (2015), https://doi.org/10.61092/iaea.3dbe-055p

[7] INTERNATIONAL ATOMIC ENERGY AGENCY, Advisory Material for the IAEA Regulations for the Safe Transport of Radioactive Material (2018 Edition), IAEA Safety Standards Series No. SSG-26 (Rev. 1), IAEA, Vienna (2022), https://doi.org/10.61092/iaea.qz7d-jiym

[8] INTERNATIONAL ATOMIC ENERGY AGENCY, INTERNATIONAL CIVIL AVIATION ORGANIZATION, INTERNATIONAL MARITIME ORGANIZATION, Preparedness and Response for a Nuclear or Radiological Emergency Involving the Transport of Radioactive Material, IAEA Safety Standards Series No. SSG-65, IAEA, Vienna (2022).

[9] INTERNATIONAL ATOMIC ENERGY AGENCY, Compliance Assurance for the Safe Transport of Radioactive Material, IAEA Safety Standards Series No. SSG-78, IAEA, Vienna (2023).

[10] INTERNATIONAL ATOMIC ENERGY AGENCY, The Management System for the Safe Transport of Radioactive Material, IAEA Safety Standards Series No. TS-G-1.4, IAEA, Vienna (2008).

[11] INTERNATIONAL ATOMIC ENERGY AGENCY, Radiation Protection Programmes for the Transport of Radioactive Material, IAEA Safety Standards Series No. SSG-86, IAEA, Vienna (2023).

[12] INTERNATIONAL ATOMIC ENERGY AGENCY, Schedules of Provisions of the IAEA Regulations for the Safe Transport of Radioactive Material (2018 Edition), IAEA Safety Standards Series No. SSG-33 (Rev. 1), IAEA, Vienna (2021).

[13] INTERNATIONAL ATOMIC ENERGY AGENCY, Format and Content of the Package Design Safety Report for the Transport of Radioactive Material, IAEA Safety Standards Series No. SSG-66, IAEA, Vienna (2022).

[14] FOOD AND AGRICULTURE ORGANIZATION OF THE UNITED NATIONS, INTERNATIONAL ATOMIC ENERGY AGENCY, INTERNATIONAL LABOUR OFFICE, PAN AMERICAN HEALTH ORGANIZATION, WORLD HEALTH ORGANIZATION, Criteria for Use in Preparedness and Response for a Nuclear or Radiological Emergency, IAEA Safety Standards Series No. GSG-2, IAEA, Vienna (2011). (A revision of this publication is in preparation.)

[15] FOOD AND AGRICULTURE ORGANIZATION OF THE UNITED NATIONS, INTERNATIONAL ATOMIC ENERGY AGENCY, INTERNATIONAL LABOUR OFFICE, PAN AMERICAN HEALTH ORGANIZATION, UNITED NATIONS OFFICE FOR THE COORDINATION OF HUMANITARIAN AFFAIRS, WORLD HEALTH ORGANIZATION, Arrangements for Preparedness for a Nuclear or Radiological Emergency, IAEA Safety Standards Series No. GS-G-2.1, IAEA, Vienna (2007). (A revision of this publication is in preparation.)

[16] FOOD AND AGRICULTURE ORGANIZATION OF THE UNITED NATIONS, INTERNATIONAL ATOMIC ENERGY AGENCY, INTERNATIONAL CIVIL AVIATION ORGANIZATION, INTERNATIONAL LABOUR OFFICE, INTERNATIONAL MARITIME ORGANIZATION, INTERPOL, OECD NUCLEAR ENERGY AGENCY, UNITED NATIONS OFFICE FOR THE COORDINATION OF HUMANITARIAN AFFAIRS, WORLD HEALTH ORGANIZATION, WORLD METEOROLOGICAL ORGANIZATION, Arrangements for the Termination of a Nuclear or Radiological Emergency, IAEA Safety Standards Series No. GSG-11, IAEA, Vienna (2018).

[17] INTERNATIONAL ORGANIZATION FOR STANDARDIZATION, Nuclear Energy — Packagings for the Transport of Uranium Hexafluoride (UF_6), ISO 7195:2020, ISO, Geneva (2020).

[18] INTERNATIONAL MARITIME ORGANIZATION, International Maritime Dangerous Goods (IMDG) Code, IMO, London (2022).

[19] INTERNATIONAL ORGANIZATION FOR STANDARDIZATION, Radiation Protection — Sealed Sources — Leakage Test Methods, ISO 9978:2020, ISO, Geneva (2020).

[20] UNITED NATIONS, Recommendations on the Transport of Dangerous Goods, Model Regulations, ST/SG/AC.10/1/Rev.23, 2 vols, United Nations, New York and Geneva (2023),
https://doi.org/10.18356/9789210019057c002

[21] INTERNATIONAL ORGANIZATION FOR STANDARDIZATION, Series 1 Freight Containers — Specification and Testing — Part 1: General Cargo Containers for General Purposes, ISO 1496-1:2013, ISO, Geneva (2013) and subsequent Amendments 1:2016, 2:2024.

[22] INTERNATIONAL ORGANIZATION FOR STANDARDIZATION, Radiological Protection — Sealed Radioactive Sources — General Requirements and Classification, ISO 2919:2012, ISO, Geneva (2012).

Annex I

SUMMARY OF APPROVAL AND PRIOR NOTIFICATION REQUIREMENTS

This summary reflects the contents of the Regulations for the Safe Transport of Radioactive Material (2025 Edition). The user's attention is called to the fact that there may be deviations (e.g. exceptions, additions) relative to:

(a) National regulations relating to safety;
(b) *Carrier* restrictions;
(c) National regulations relating to nuclear security, physical protection, liability, insurance, pre-notification and/or routeing and import/export/transit licensing.[1]

[1] In particular, additional measures are taken to provide appropriate physical protection in the transport of nuclear material and to prevent acts without lawful authority that constitute the receipt, possession, use, transfer, alteration, disposal or dispersal of nuclear material and which cause, or are likely to cause, death or serious injury to any person or substantial damage to property or to the environment (see Refs [I–1] to [I–10]).

ANNEX I: SUMMARY OF APPROVAL AND PRIOR NOTIFICATION REQUIREMENTS (Part 1)

Key paragraphs in the Regulations	Class of *package* or material	*Competent authority approval* required		*Consignor* required to notify country of origin and countries en route[a] of each *shipment*
		Country of origin	Countries en route[a]	
	Excepted package[b,c]	No	No	No
	LSA material[c,d,e] and *SCO-I*[c,e] and *SCO-II*[c,e] — *Type IP-1*, — *Type IP-2* or — *Type IP-3*	No	No	No
	Type A[c,d,e]	No	No	No
520, 825, 826	*SCO-III* — *Shipment*	Yes	Yes	No

[a] Countries *through or into* which (but not over which) the *consignment* is transported (see para. 204 of the Regulations).

[b] For international transport by post, the *consignment* shall be deposited with the postal service only by *consignors* authorized by the national authority.

[c] If the *radioactive contents* are *fissile material* excepted under para. 417(f) of the Regulations, *multilateral approval* shall be required (see para. 805 of the Regulations).

[d] If the *radioactive contents* are uranium hexafluoride in quantities of 0.1 kg or more, the *approval* requirements for *packages* containing it shall additionally apply (see paras 802 and 807 of the Regulations).

[e] If the *radioactive contents* are *fissile material* that is not excepted from the requirements for *packages* containing *fissile material*, then the *approval* requirements in paras 814, 825 and 826 of the Regulations shall additionally apply.

ANNEX I: SUMMARY OF APPROVAL AND PRIOR NOTIFICATION REQUIREMENTS (Part 2)

Key paragraphs in the Regulations	Class of *package* or material	*Competent authority approval* required		*Consignor* required to notify country of origin and countries en route[a] of each *shipment*
		Country of origin	Countries en route[a]	
	Type B(U)[b,c,d]			
808	— *Package design*	Yes	No[e]	
559, 560, 825, 826	— *Shipment*	No	No	(see Notes 1 and 2)
	Type B(M)[b,c,e]			
811	— *Package design*	Yes	Yes	Yes
559, 560, 825, 826	— *Shipment*	(see Note 3)	(see Note 3)	(see Note 1)
	Type C [b,c,d]			
808	— *Package design*	Yes	No	
559, 560, 825, 826	— *Shipment*	No	No	(see Notes 1 and 2)

[a] Countries *through or into* which (but not over which) the *consignment* is transported (see para. 204 of the Regulations).

[b] If the *radioactive contents* are *fissile material* that is not excepted from the requirements for *packages* containing *fissile material*, then the *approval* requirements in paras 814, 825 and 826 of the Regulations shall additionally apply.

[c] If the *radioactive contents* are uranium hexafluoride in quantities of 0.1 kg or more, the *approval* requirements for *packages* containing it shall additionally apply (see paras 802 and 807 of the Regulations).

[d] If the *radioactive contents* are *fissile material* excepted under para. 417(f) of the Regulations, *multilateral approval* shall be required (see para. 805 of the Regulations).

[e] If the *radioactive contents* are *low dispersible radioactive material* and the *package* is to be shipped by air, *multilateral approval* of the *package design* is required (see para. 808(b) of the Regulations).

Note 1: Before the first *shipment* of a *package* requiring *competent authority approval* of the *design*, the *consignor* shall ensure that a copy of the certificate of *approval* for that *design* has been submitted to the *competent authority* of each country (see para. 557 of the Regulations).

Note 2: Notification is required if the *radioactive contents* exceed $3000A_1$ or $3000A_2$ or 1000 TBq, whichever is the lower (see para. 558 of the Regulations).

Note 3: *Multilateral approval* of *shipment* required if the *radioactive contents* exceed $3000A_1$ or $3000A_2$ or 1000 TBq, whichever is the lower, or if controlled intermittent venting is allowed (see paras 825 and 826 of the Regulations).

ANNEX I: SUMMARY OF APPROVAL AND PRIOR NOTIFICATION REQUIREMENTS (Part 3)

Key paragraphs in the Regulations	Class of *package* or material	*Competent authority approval* required		*Consignor* required to notify country of origin and countries en route[a] of each *shipment*
		Country of origin	Countries en route[a]	
	Packages for *fissile material*			
814	— *Package design*	Yes[b]	Yes[b]	
825, 826	— *Shipment*			
	$\Sigma CSI \leq 50$	No	No	(see Notes 1 and 2)
	$\Sigma CSI > 50$	Yes	Yes	(see Notes 1 and 2)
	Packages containing 0.1 kg or more of uranium hexafluoride[c]			
807	— *Package design*	Yes	Yes for H(M), no for H(U)	
825, 826	— *Shipment*	No[d]	No[d]	(see Notes 1 and 2)

[a] Countries *through or into* which (but not over which) the *consignment* is transported (see para. 204 of the Regulations).

[b] *Designs* of *packages* containing *fissile material* may also require *approval* in respect of one of the other items in Annex I.

[c] If the *radioactive contents* are *fissile material* excepted under para. 417(f) of the Regulations, *multilateral approval* shall be required (see para. 805 of the Regulations).

[d] *Shipments* may, however, require *approval* in respect of one of the other items in Annex I.

Note 1: The *multilateral approval* requirement for *packages* containing *fissile material,* and for some *packages* containing uranium hexafluoride, automatically satisfies the requirement of para. 557 of the Regulations.

Note 2: Notification is required if the *radioactive contents* exceed $3000A_1$ or $3000A_2$ or 1000 TBq, whichever is the lower (see para. 558 of the Regulations).

ANNEX I: SUMMARY OF APPROVAL AND PRIOR NOTIFICATION REQUIREMENTS (Part 4)

Key paragraphs in the Regulations	Class of *package* or material	*Competent authority approval* required		*Consignor* required to notify country of origin and countries en route[a] of each *shipment*
		Country of origin	Countries en route[a]	
	Special form radioactive material			
803	— *Design*	Yes	No	No
825, 826	— *Shipment*	(see Note 1)	(see Note 1)	(see Note 1)
	Low dispersible radioactive material			
803	— *Design*	Yes	Yes	No
825, 826	— *Shipment*	(see Note 1)	(see Note 1)	(see Note 1)
	Special arrangement			
560, 802, 831	— *Shipment*	Yes	Yes	Yes
	Type B(U) packages for which *design* is approved under			
820	— 1985 Regulations	Yes	Yes	(see Note 2)
805	*Fissile material* excepted from "FISSILE" classification, in accordance with para. 606	Yes	Yes	No
817	Exempt *consignment* of instruments or articles	Yes	Yes	No

[a] Countries *through or into* which (but not over which) the *consignment* is transported (see para. 204 of the Regulations).

Note 1: See *approval* and prior notification requirements for the applicable *package*.

Note 2: Before the first *shipment* of a *package* requiring *competent authority approval* of the *design*, the *consignor* shall ensure that a copy of the certificate of *approval* for that *design* has been submitted to the *competent authority* of each country (see para. 557 of the Regulations).

REFERENCES TO ANNEX I

[I–1] Convention on the Physical Protection of Nuclear Material, INFCIRC/274/Rev.1, IAEA, Vienna (1980).

[I–2] Amendment to the Convention on the Physical Protection of Nuclear Material, INFCIRC/274/Rev.1/Mod.1 (Corrected), IAEA, Vienna (2021).

[I–3] INTERNATIONAL ATOMIC ENERGY AGENCY, Nuclear Security Recommendations on Physical Protection of Nuclear Material and Nuclear Facilities (INFCIRC/225/Revision 5), IAEA Nuclear Security Series No. 13, IAEA, Vienna (2011),
https://doi.org/10.61092/iaea.ko2c-dc4q

[I–4] INTERNATIONAL ATOMIC ENERGY AGENCY, Physical Protection of Nuclear Material and Nuclear Facilities (Implementation of INFCIRC/225/Revision 5), IAEA Nuclear Security Series No. 27-G, IAEA, Vienna (2018).

[I–5] INTERNATIONAL ATOMIC ENERGY AGENCY, Security of Radioactive Material in Transport, IAEA Nuclear Security Series No. 9-G (Rev. 1), IAEA, Vienna (2020).

[I–6] INTERNATIONAL ATOMIC ENERGY AGENCY, Nuclear Security Recommendations on Radioactive Material and Associated Facilities, IAEA Nuclear Security Series No. 14, IAEA, Vienna (2011).

[I–7] INTERNATIONAL ATOMIC ENERGY AGENCY, Code of Conduct on the Safety and Security of Radioactive Sources, IAEA, Vienna (2004).

[I–8] INTERNATIONAL ATOMIC ENERGY AGENCY, Guidance on the Import and Export of Radioactive Sources, IAEA, Vienna (2012).

[I–9] INTERNATIONAL ATOMIC ENERGY AGENCY, Security of Nuclear Material in Transport, IAEA Nuclear Security Series No. 26-G, IAEA, Vienna (2015).

[I–10] INTERNATIONAL ATOMIC ENERGY AGENCY, Security of Nuclear and Other Radioactive Material in Transport, IAEA Nuclear Security Series No. 46-T, IAEA, Vienna (2024),
https://doi.org/10.61092/iaea.4umu-2uji

Annex II

PREFIXES AND SYMBOLS OF THE INTERNATIONAL SYSTEM OF UNITS

This edition of the Regulations uses the International System of Units (SI). The decimal multiples and submultiples of a unit may be formed by prefixes or symbols, having the following meanings, placed before the name or symbol of the unit:

Multiplying factor	Prefix	Symbol
1 000 000 000 000 000 000 = 10^{18}	exa	E
1 000 000 000 000 000 = 10^{15}	peta	P
1 000 000 000 000 = 10^{12}	tera	T
1 000 000 000 = 10^{9}	giga	G
1 000 000 = 10^{6}	mega	M
1 000 = 10^{3}	kilo	k
100 = 10^{2}	hecto	h
10 = 10^{1}	deca	da
0.1 = 10^{-1}	deci	d
0.01 = 10^{-2}	centi	c
0.001 = 10^{-3}	milli	m
0.000 001 = 10^{-6}	micro	μ
0.000 000 001 = 10^{-9}	nano	n
0.000 000 000 001 = 10^{-12}	pico	p
0.000 000 000 000 001 = 10^{-15}	femto	f
0.000 000 000 000 000 001 = 10^{-18}	atto	a

Annex III

SUMMARY OF CONVEYANCES AND LARGE FREIGHT CONTAINERS REQUIRED TO BE UNDER EXCLUSIVE USE

Conveyances and *large freight containers* are required to be under *exclusive use* when loaded with any of the following:

(a) Unpackaged *LSA-I material*, *SCO-I* and *SCO-III* (see para. 520);
(b) Liquid *LSA-I material* in a *Type IP-1 package* (see para. 521 and Table 5);
(c) Gaseous and/or liquid *LSA-II material* in a *Type IP-2 package* (see para. 521 and Table 5);
(d) *LSA-III material* in a *Type IP-2 package* (see para. 521 and Table 5);
(e) *Packages,* other than *freight containers* used as *packaging*, or *overpacks* with an individual *TI* greater than 10 (see para. 526(a));
(f) *Freight containers* used as *packaging* with an individual *TI* greater than 50 (see para. 526(b));
(g) *Packages* or *overpacks* having the maximum *dose rate* at any point on the external surfaces that exceeds 2 mSv/h (see para. 527);
(h) *Packages*, *overpacks* and *freight containers* with a total sum of *TI* exceeding the values given in Table 10 (see para. 566(a));
(i) *Packages, overpacks* and *freight containers* with a total sum of *CSI* exceeding the values given in Table 11 for "not under *exclusive use*" (see paras 526(c), 566(c) and 569);
(j) *Type B(U), Type B(M)* or *Type C package* whose temperature of accessible surfaces exceeds 50°C when subject to an ambient temperature of 38°C in the absence of insolation (see para. 654);
(k) Up to 45 g of *fissile nuclides* on a *conveyance*, either packaged or unpackaged, in accordance with the provisions of paras 417(e) and 570(e);
(l) *Packages* containing *fissile material* classified as non-fissile or fissile-excepted under para. 417(a)(i) or (iii) of the 2009 Edition of these Regulations (see para. 822).

CONTRIBUTORS TO DRAFTING AND REVIEW (2025)

Abe, M.	NFT Corporation, Japan
Ahart, R.	International Atomic Energy Agency
Al-Hajri, R.	Ministry of Municipality and Environment, Qatar
Al Yammahi, A.	Federal Authority for Nuclear Regulation, United Arab Emirates
Alvano, P.	National Inspectorate for Nuclear Safety and Radiation Protection, Italy
Ashtiani, H.K.	Iran Nuclear Regulatory Authority, Islamic Republic of Iran
Barki, A.	Federal Ministry for Climate Action, Environment, Energy, Mobility, Innovation and Technology, Austria
Barto, A.	Nuclear Regulatory Commission, United States of America
Bertreix, P.	European Organization for Nuclear Research, Switzerland
Boyle, R.	Department of Transportation, United States of America
Brennan, D.	International Air Transport Association
Brennes, M.E.	Ministry of Public Works and Transport, Costa Rica
Bruno, N.C.	National Nuclear Energy Commission, Brazil
Buchelnikov, A.	State Atomic Energy Corporation "Rosatom", Russian Federation
Bujnová, A.	International Atomic Energy Agency
Cabianca, T.	UK Health Security Agency, United Kingdom
Cao, F.F.	Nuclear and Radiation Safety Centre, China

CONTRIBUTORS TO DRAFTING AND REVIEW (2025)

Chan, C.	International Air Transport Association
Charette, M.A.	Cameco Corporation, Canada
Chavan, V.	Bhabha Atomic Research Centre, India
Chichinadze, D.	Agency of Nuclear and Radiation Safety, Georgia
Chrupek, T.	Authority for Nuclear Safety and Radiation Protection, France
Clark, S.	Office for Nuclear Regulation, United Kingdom
Collard, O.	European Commission
Crane, S.	World Nuclear Transport Institute
Daigle, D.	Transport Canada, Canada
Darras, F.	International Organization for Standardization
Desiderio, V.	Universal Postal Union
Desnoyers, B.	International Organization for Standardization
Dessent, K.	Australian Radiation Protection and Nuclear Safety Agency, Australia
Duffy, J.	Environmental Protection Agency, Ireland
Dwyer, R.	Canadian Nuclear Safety Commission, Canada
Eberhardt, H.	GRS, Germany
Ecevit, S.T.	Nuclear Regulatory Authority, Türkiye
Edwards, W.S.	World Nuclear Transport Institute
Ejaz, M.U.	International Atomic Energy Agency
Elechosa, C.	Nuclear Regulatory Authority, Argentina
Endres, A.	Federal Ministry for the Environment, Climate Action, Nature Conservation and Nuclear Safety, Germany
Ershov, V.	State Atomic Energy Corporation “Rosatom”, Russian Federation
Fang, X.B.	National Nuclear Safety Administration, China

CONTRIBUTORS TO DRAFTING AND REVIEW (2025)

Fayyaz, S.	International Atomic Energy Agency
Féron, F.	Authority for Nuclear Safety and Radiation Protection, France
Fisher, D.	World Nuclear Transport Institute
Freeman, E.	Ontario Power Generation, Canada
Frosio, T.	European Organization for Nuclear Research, Switzerland
Fukuda, T.	Nuclear Regulation Authority, Japan
Fulford, G.	International Source Suppliers and Producers Association
Furuta, M.	Nuclear Regulation Authority, Japan
García, M.L.	Nuclear Safety Council, Spain
Gauthier, F.	Authority for Nuclear Safety and Radiation Protection, France
Giannone, B.	National Inspectorate for Nuclear Safety and Radiation Protection, Italy
Gopal-Kali, M.	Authority for Nuclear Safety and Radiation Protection, Kingdom of the Netherlands
Gorlin, S.	World Nuclear Association
Grozev, N.	Nuclear Regulatory Agency, Bulgaria
Hajizadeh, B.	Iran Nuclear Regulatory Authority, Islamic Republic of Iran
Hellstèn, S.	Radiation and Nuclear Safety Authority, Finland
Hinrichsen, P.	National Nuclear Regulator, South Africa
Hirose, M.	National Maritime Research Institute, Japan
Holländer, L.	Federal Office for the Safety of Nuclear Waste Management, Germany
Hubert, R.	United Nations Economic Commission for Europe

CONTRIBUTORS TO DRAFTING AND REVIEW (2025)

Hughes-Gill, K.	Office for Nuclear Regulation, United Kingdom
Ito, D.	World Nuclear Transport Institute
Jakowiuk, A.	National Atomic Energy Agency, Poland
Jones, C.	Office for Nuclear Regulation, United Kingdom
Katona, T.	Hungarian Atomic Energy Authority, Hungary
Kattner, F.	World Nuclear Transport Institute
Kim, H.S.	Korea Institute of Nuclear Safety, Republic of Korea
Kirkin, A.	Scientific and Engineering Centre for Nuclear and Radiation Safety, Russian Federation
Klimo, P.	Nuclear Regulatory Authority of the Slovak Republic, Slovakia
Koch, F.	Swiss Federal Nuclear Safety Inspectorate, Switzerland
Komann, S.	Federal Institute for Materials Research and Testing, Germany
Konnai, A.	National Maritime Research Institute, Japan
Kontarovich, S.	Israel Atomic Energy Commission, Israel
Kruehler, J.	World Nuclear Transport Institute
Krüger, A.C.	Danish Health Authority, Denmark
Kuriyama, K.	National Maritime Research Institute, Japan
Lacasse, S.	BWXT Nuclear Energy Canada Inc., Canada
Lamarche, D.	Transport Canada, Canada
Lawrence, C.	Australian Radiation Protection and Nuclear Safety Agency, Australia
Ledroit, F.	Authority for Nuclear Safety and Radiation Protection, France
Lee, B.H.	Korea Institute of Nuclear Safety, Republic of Korea

Leekha, R.	Ontario Power Generation, Canada
Lemoine, E.	Canadian Nuclear Safety Commission, Canada
Lempiäinen, T.	International Federation of Air Line Pilots' Associations
Li, G.Q.	China Institute for Radiation Protection, China
Louis, B.	Authority for Nuclear Safety and Radiation Protection, France
Lourtie, G.	Federal Agency for Nuclear Control, Belgium
Lynchahon, K.	Ontario Power Generation, Canada
Mahlert, C.	Authority for Nuclear Safety and Radiation Protection, Kingdom of the Netherlands
Malesys, P.	World Nuclear Transport Institute
Manjunath, B.S.	Bhabha Atomic Research Centre, India
Margerison, I.	Tantalum-Niobium International Study Center
Mashimo, K.	Nuclear Regulation Authority, Japan
Matos, L.V.S.	National Nuclear Energy Commission, Brazil
Mazrou, H.	Nuclear Research Center of Algiers, Algeria
Mennerdahl, D.	E Mennerdahl Systems, Sweden
Miller, J.J.	International Source Suppliers and Producers Association
Mireles, A.	United Nations Economic Commission for Europe
Molatim, T.Z.	Libyan Atomic Energy Establishment, Libya
Moussa, J.	Norwegian Radiation and Nuclear Safety Authority, Norway
Nahar, M.	Bangladesh Atomic Energy Regulatory Authority, Bangladesh
Nissaka, H.	Nuclear Regulation Authority, Japan

Nitsche, F.	Consultant, Germany
O'Connor, G.	Office for Nuclear Regulation, United Kingdom
Ozaki, K.	Nuclear Regulation Authority, Japan
Penda, E.	World Nuclear Transport Institute
Peng, A.C.	National Environment Agency, Singapore
Persico, D.	Tantalum-Niobium International Study Center
Pilecki, L.	Canadian Nuclear Safety Commission, Canada
Pilgrim, D.	Canadian Nuclear Laboratories, Canada
Pourhatam, A.	Iran Nuclear Regulatory Authority, Islamic Republic of Iran
Pourmehr, M.	Iran Nuclear Regulatory Authority, Islamic Republic of Iran
Pstrak, D.	Nuclear Regulatory Commission, United States of America
Ramsay, J.	Canadian Nuclear Safety Commission, Canada
Rathje-Unger, L.	Federal Ministry for Digital and Transport, Germany
Reber, E.	International Atomic Energy Agency
Reculeau, J.	Defence Nuclear Safety Authority, France
Reiche, I.	Federal Office for the Safety of Nuclear Waste Management, Germany
Rolencík, M.	Nuclear Regulatory Authority of the Slovak Republic, Slovakia
Rooney, K.	International Civil Aviation Organization
Ruprecht, B.	Federal Office for the Safety of Nuclear Waste Management, Germany
Sarkar, S.	Australian Radiation Protection and Nuclear Safety Agency, Australia

Schulze Grachtrup, D.	Federal Office for the Safety of Nuclear Waste Management, Germany
Schwela, U.	Tantalum-Niobium International Study Center
Shaw, P.	International Atomic Energy Agency
Shima, M.	Nuclear Regulation Authority, Japan
Smith, J.	Nuclear Regulatory Commission, United States of America
Sollund, H.	Norwegian Radiation and Nuclear Safety Authority, Norway
Song, B.	International Maritime Organization
Stav, K.	Israel Atomic Energy Commission, Israel
Steiner, C.	Federal Ministry for Climate Action, Environment, Energy, Mobility, Innovation and Technology, Austria
Suter, P.	Axpo Power AG, Switzerland
Takkienen, A.	Radiation and Nuclear Safety Authority, Finland
Ter Morshuizen, M.G.	Authority for Nuclear Safety and Radiation Protection, Kingdom of the Netherlands
Thomas, C.	World Nuclear Transport Institute
Trivelloni, S.	National Inspectorate for Nuclear Safety and Radiation Protection, Italy
Václav, J.	Nuclear Regulatory Authority of the Slovak Republic, Slovakia
Vogiatzi, S.	Greek Atomic Energy Commission, Greece
Wallin, M.	Swedish Radiation Safety Authority, Sweden
Wille, F.	Federal Institute for Materials Research and Testing, Germany
Won, J.K.	Korea Institute of Nuclear Safety, Republic of Korea
Wright, R.	International Atomic Energy Agency

Zahn, Y.L.	China National Nuclear Corporation, China
Zamora, F.M.	Nuclear Safety Council, Spain

Numerous other participants in Member States took part in the review and revision of this publication. Their invaluable contribution to the process is recognized.

INDEX

(by paragraph number)

IAEA
International Atomic Energy Agency

CONTACT IAEA PUBLISHING

Feedback on IAEA publications may be given via the on-line form available at:
www.iaea.org/publications/feedback

This form may also be used to report safety issues or environmental queries concerning IAEA publications.

Alternatively, contact IAEA Publishing:

Publishing Section
International Atomic Energy Agency
Vienna International Centre, PO Box 100, 1400 Vienna, Austria
Telephone: +43 1 2600 22529 or 22530
Email: sales.publications@iaea.org
www.iaea.org/publications

Priced and unpriced IAEA publications may be ordered directly from the IAEA.

ORDERING LOCALLY

Priced IAEA publications may be purchased from regional distributors and from major local booksellers.

FSC
www.fsc.org
MIX
Papier | Fördert
gute Waldnutzung
FSC® C083411

Zeitfracht Medien GmbH
Ferdinand-Jühlke-Straße 7
99095 Erfurt, Deutschland
produktsicherheit@kolibri360.de